T0136305

Hybrid Encryption Algorithms Over Wireless Communication Channels

Mai Helmy Shaheen

CRC Press is an imprint of the
Taylor & Francis Group, an **informa** business
A SCIENCE PUBLISHERS BOOK

First edition published 2021
by CRC Press
6000 Broken Sound Parkway NW, Suite 300, Boca Raton, FL 33487-2742

and by CRC Press
2 Park Square, Milton Park, Abingdon, Oxon, OX14 4RN

CRC Press is an imprint of Taylor & Francis Group, LLC

ISBN: 9780367508159 (hbk)

Typeset in Palatino
by Radiant Productions

Preface

We try in this book to look it image and video encryption with the eyes of communication researches. Traditional studies of encryption concentrate on the strength of the encryption algorithm without taking into consideration what is after encryption. What is after encryption is the question we must answer to select the appropriate encryption algorithm. For real-life applications, what is after encryption is communication of encrypted images and videos. With the advances in mobile and TV applications, we have to transmit encrypted images and videos wirelessly. So, "Do our encryption algorithms tolerate the wireless communication impairments?", that is the question we are trying to answer in this book.

We can summarize the main contributions in this book as:

1. This book is devoted to the issue of images and video encryption for the purpose of wireless communications.
2. Diffusion as well as permutation ciphers are considered in this book, with a comparison between them using different evaluation metrics.
3. Modifications are presented to existing block ciphers either to speed or enhance their performance.
4. The wireless communication environment, in which the encrypted images and videos need to be communicated, is studied.
5. Simulation experiments are presented for the validation of the discussed algorithms and modifications, and also for investigating the performance of algorithms over wireless channels.
6. MATLAB® codes for most of the simulation experiments in this book are included in two appendices at the end of the book.

Finally, we hope that this book will be helpful for the images and video processing, and wireless communication communities.

Acknowledgements

ALLAH is the first and the last to be thanked.

I want to thank all who helped me by their knowledge and experience. I will always appreciate their efforts. I'm extremely grateful to Prof. Ibrahim Eldokany for his valuable suggestions and support. I'm also grateful to Prof. El-Sayed El-Rabaie for his valuable discussions and support. I would like to thanks Prof. Fathi Abd El-samie for his supervision and continuous support and helpful discussions throughout this work and the time and effort he spent for my work.

My sincere appreciation and gratitude are devoted to my parents, my sister, and my brothers for their help and patience during the preparation of this work.

Contents

List of Abbreviations

ACI	Adjacent Channel Interference
A/D	Analog to Digital
AES	Advanced Encryption Standard
AM	Amplitude Modulation
ASK	Amplitude Shift Keying
AWGN	Additive White Gaussian Noise
BER	Bit Error Rate
CBC	Cipher Block Chaining
CCFD	Complementary Cumulative Distribution Function
CFD	Cumulative Distribution Function
CFB	Cipher Feedback
CFO	Carrier Frequency Offset
CR	Clipping Ratio
D/A	Digital to Analog
DCT-OFDM	Discrete Cosine Transform based OFDM
DES	Data Encryption Standard
DRPE	Double Random Phase Encoding
DSL	Digital subscriber lines
3DV	3-D Video
DVB	Digital Video Broadcasting
DVB-C	DVB including cable systems
DVB-S	DVB including satellite
DVB-T	DVB including terrestrial transmission
DWT-OFDM	Discrete Wavelet Transform based OFDM
ECB	Electronic Code Book
EER	Equal Error Rate
FDM	Frequency Division Multiplexing
FEC	Forward Error Correction
FFT-OFDM	Fast Fourier Transform based OFDM
F-Function	Feistel function

FM	Frequency Modulation
FSK	Frequency Shift Keying
GF	Galois Field
HDTV	High Definition Television
HGI	Hybrid Guard Interval
HIPERLAN2	High Performance Local Area Network
IBM	International Business Machines
ICI	Inter-Carrier Interference
ID	Identification
IDCT	Inverse Discrete Cosine Transform
IDWT	Inverse Discrete Wavelet Transform
IEEE	Institute of Electrical and Electronics Engineers
IFFT	Inverse Fast Fourier Transformer
IP	Initial Permutation
ISI	Intersymbol Interference
IV	initial vector
MCM	Multicarrier modulation
MIC	Multi-view Image Compression
MVC	Multi-view Video Compression
MPEG	Motion Picture Experts Group
MSE	Mean Square Error
NIST	National Institute of Standards and Technology
OBL	Outside-Broadcasting-Link
OFB	Output Feedback
OFDM	Orthogonal Frequency Division Multiplexing
PAPR	Peak to Average Power Ratio
Pdf	Probability density function
PFD	Probability of False Distribution
PINs	Personal Identification Numbers
PLC	Power line communication
PM	Phase Modulation
PSE	Pilot Symbol Estimation
PSK	Phase shift keying
PSNR	Peak Signal-to-Noise Ratio
PTD	Probability of True Distribution
QAM	Quadrature Amplitude Modulation
RC6	Ron's Code six
RSA	Rivest Shamir Adelman
Si	Substitution boxes

S	Block Size
SC	Self-Cancellation
SIFT	Scale-Invariant Feature Transform
SFN	Single Frequency Network
TDM	Time Division Multiplexing
UHF	Ultra High Frequency
WLANs	Wireless local area networks
ZF	Zero forcing
ZP	Zero Padding

Chapter 1

Introduction

The demand for high speed and efficient wireless communication systems has grown in the last few decades. The progress in mobile communications, satellite applications, internet applications, and computer networks has given rise to new problems with regard to security and privacy. Having a secure and reliable means of communicating with images and video is becoming a necessity, and the related issues must be carefully considered. Hence, network security and data encryption have become important. Images can be considered, nowadays, as one of the most usable forms of information. Image and video encryption have applications in various fields, including internet communications, multimedia systems, medical imaging, telemedicine, and military communications. There are two main types of applications responsible for information transmission in communication systems. The first one considers speed as the main issue, as it deals with online processing. In this type of application, the encryption process must be performed very fast, even if the security of the encryption algorithm is not powerful enough. The second one considers security as the main issue. In this type of application, data must be encrypted using highly-secure encryption algorithms [1, 2].

The objective of this book is to study the transmission of encrypted images over wireless channels with OFDM. Several encryption algorithms have been investigated in the book. Also, the different versions of OFDM, such as Fast Fourier Transform-based OFDM, Discrete Cosine Transform-based OFDM, and Discrete Wavelet Transform-based OFDM, are studied. The objective of this study is to make a trade-off between the encryption algorithm and the OFDM version that should be used for the best performance in image and video transmission for both Additive White Gaussian Noise (AWGN) and fading channels [2, 4].

1.1 Processing of Encrypted Data

The goal of third and fourth generation mobile networks is to provide users with high data rates, and to provide a wider range of services, such as voice communications, videophones, and high-speed Internet access. A common challenge in designing a wireless system is to overcome the effects of the wireless channel, which is characterized as having multiple transmission paths and as being time varying. OFDM has a promising future as a new technology in several next generation wireless communication systems. The ability of OFDM systems to combat the effects of multipath propagation with a comparatively simple receiver structure made it the modulation of choice for some of the most prominent wireless technologies such as the Institute of Electrical and Electronics Engineers (IEEE 802.11) Wireless Local Area Networks (WLANs). It is also used in wireless broadcasting applications such as Digital Audio Broadcasting (DAB) and terrestrial Digital Video Broadcasting (DVB-T) [5]. OFDM has been also implemented in wireline applications such as Digital Subscriber Lines (DSL) and Power Line Communication (PLC) [6, 7].

In a conventional serial data system, the symbols are transmitted sequentially, and the frequency spectrum of each data symbol is allowed to occupy the entire available bandwidth. In a parallel data transmission system, several symbols are transmitted at the same time, and this offers the possibility of alleviating many of the problems encountered in serial systems. In OFDM, the data is divided among many closely-spaced carriers. This accounts for the frequency division multiplexing part of the name. This is not a multiple access technique, since there is no common medium to be shared. The entire bandwidth is filled from a single source of data. Instead of transmitting in a serial way, data is transferred in a parallel way. OFDM can be simply defined as a form of multi-carrier modulation, in which carrier spacing is carefully selected so that each sub-carrier is orthogonal to the other sub-carriers. These orthogonal signals can be separated at the receiver. Orthogonality can be achieved by carefully selecting the carrier spacing by letting the carrier spacing be equal to the reciprocal of the useful symbol period, as explained later [7].

OFDM communication systems have two primary drawbacks. The *first* is the high sensitivity to carrier frequency offsets and phase noise. When there are frequency offsets in the sub-carriers, the orthogonality among the sub-carriers breaks and this causes Inter-Carrier Interference (ICI). The second drawback is that the transmitted OFDM signal has large amplitude fluctuations, and so a high Peak-to-Average Power Ratio (PAPR). This high PAPR requires system components with a wide linear range in order to accommodate for the signal variations. Nonlinear distortion occurs as a result of any loss of sub-carrier orthogonality, and hence a degradation in the system performance occurs.

1.2 Objectives of the Book

This book presents a framework for transmission of Rubik's cube encrypted images and videos instead of normal images and videos. Encryption adds a degree of security as compared to traditional image communication schemes.

Transmission of images and videos in different communication systems is a very important topic that has attracted the attention of researchers in the last few decades. The endeavor towards an optimum encryption scheme and an optimum modulation method for this purpose is still under consideration. Orthogonal frequency division multiplexing (OFDM) is an attractive multicarrier transmission technique for wideband communications, because it effectively transforms the frequency selective fading channel into a flat fading channel. Hence, OFDM provides greater immunity to multipath fading and impulsive noise and eliminates the need for complicated equalizers. However, OFDM communication systems suffer from fading effects, high sensitivity to carrier frequency offsets and phase noise.

In this book, we present a study for two different families of encryption schemes: permutation-based schemes and diffusion-based schemes. The objective of this study is to select the more suitable scheme for encrypted image transmission over communication systems. Encryption and decryption quality metrics are used in the presented comparison. Two modifications to enhance the properties of the permutation and diffusion algorithms are presented in this book. In the proposed hybrid encryption framework, we need to achieve both diffusion and permutation in the encrypted Three-Dimensional (3-D) images. Towards this target, we used chaotic, Ron's Code (RC6) or Advanced Encryption Standard (AES) technique in a pre-processing step to achieve the permutation and diffusion. The Rubik's cube is used afterwards to achieve a greater degree of permutation. Then, we studied its sensitivity to the wireless channel impairments, and the effect of channel equalization on the received 3-D images quality.

1.3 Book Organization

Chapter 2 covers the traditional permutation- and diffusion-based image encryption and transmission through OFDM system. The chapter concentrates on AES, RC6 algorithm, and chaotic Baker map algorithm.

Chapter 3 presents the Rubik's cube encryption algorithm with a comparison between this algorithm and the classical algorithms based on histogram, deviation from the original image, and immunity to noise. Also, a comparison between different OFDM versions for encrypted image transmission under different channel characteristics is presented.

In Chapter 4, a proposed hybrid image encryption algorithm for wireless communication is presented. Different modes of operation are considered in the encryption algorithms for enhanced communication performance.

Chapter 5 gives a discussion of the proposed hybrid encryption framework for reliable wireless 3-D video communication. The simulation results are also discussed in this chapter.

Finally, Chapter 6 gives the concluding remarks.

Chapter 2

Fundamentals of Image Encryption for Wireless Communications

2.1 Introduction

Fifty years ago, Claude Shannon pointed out that the fundamental techniques to encrypt a block of symbols are confusion and diffusion. Confusion can obscure the relationship between the plaintext and the ciphertext, and diffusion can spread the changes throughout the whole ciphertext. Substitution is the simplest method of confusion, and permutation is the simplest method of diffusion. Substitution replaces a symbol with another one, while permutation changes the sequence of the symbols in the block to make them unreadable. If applied independently, neither substitution nor permutation works very well [8].

These two techniques are still the foundations of encryption. In the 19th century, Kirchhoff proposed a famous theory about the security principles of any encryption system. This theory has laid the groundwork for the most important principles in designing a cryptosystem for researchers and engineers. Kirchhoff observed that the encryption algorithms are supposed to be known to the attackers. Thus, the security of an encryption system should rely on the secrecy of the encryption/decryption key instead of the encryption algorithm itself. In the very beginning, the opponent does not know the algorithm. The encryption system will not be able to protect the ciphertext once the algorithm is broken. The security level of an encryption algorithm is measured by the size of its key space. The larger the size of the key space is, the more time the attacker needs to do the exhaustive search of the key space, and thus the higher the security level is [8]. In our study, we will be concerned with three diffusion-based algorithms: the AES, and the RC6, and a permutation-based algorithm, which is the chaotic Baker map algorithm.

2.2 Encryption System Model

Encryption is a method or a process for protecting information from undesirable attacks by converting it into a form that is unrecognizable to attackers. Data encryption is mainly the scrambling of the content of the data, such as text, image, audio, video and so forth, to make the data unreadable or invisible during transmission. The inverse of encryption is data decryption, which recovers the original data.

Figure 2.1 is the general model of a typical encryption/decryption system. The encryption procedure could be described as $E\ (P, K) = C$, where P is the plaintext (original message), E is the encryption algorithm, K is the encryption key, and C is the ciphertext (scrambled message). The ciphertext is transmitted through the communication channel, which is subject to attackers. At the receiver end, the decryption procedure could be described as $D\ (C, K') = P$, where C is the ciphertext, D is the decryption algorithm, and K' is the decryption key (it may or may not be the same as the encryption key, K) [8–9].

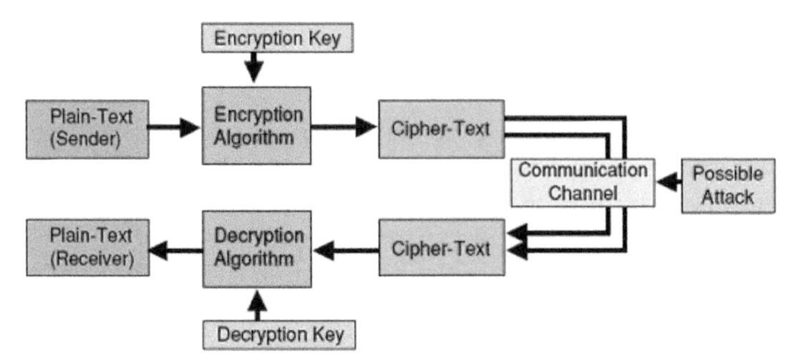

Fig. 2.1: The general model of an encryption/decryption system [9].

2.3 Key Types

There are two encryption/decryption key types: public-key and private-key.

A private-key system is also called a symmetric system, because the decryption key is the same as the encryption key. Because of its symmetric property, the encryption/decryption key must be transmitted prior to the transmission of the ciphertext. The drawback of the private-key system is that a secure communication channel for key transmission is required.

The public-key system, which is also called the asymmetric system, has a decryption key that is different from the encryption key. Each person in the group knows the encryption key. In this way, each member can use the public key to encrypt a message. Only the person, who has the decryption key, can decrypt the ciphertext. Generally, it is computationally infeasible to derive the decryption key from the encryption key, and this is how the ciphertext can be protected. With the public-key encryption system, there is no need for a secure communication channel for the transmission of the encryption key [8–9].

Depending on the type of the plaintext, data encryption systems are classified as text encryption, audio encryption, image encryption and video encryption systems. In order to have a generic cryptosystem that can encrypt digital data, such as text, image, audio, and video, some encryption standards have been developed. Among these standard, the AES and the RC6 are elaborately designed and widely adopted.

2.4 Diffusion-based Algorithms

In this book, two diffusion-based encryption algorithms are considered. They will be explained in the following sub-sections.

2.4.1 The Advanced Encryption Standard (AES)

With advanced and powerful computers, the Data Encryption Standard (DES) has proved to be insecure. As a result, in 1997, the National Institute of Standards and Technology (NIST) called for proposals for the next generation encryption standard. After three years' work, the NIST announced its selection for the AES algorithm. In 2001, the AES became the official encryption standard. It is a block-structured algorithm with variable length keys of 128 bits, 192 bits, and 256 bits. This significantly increased the security level compared to the DES [9]. The algorithm has very good performance in both hardware and software implementations. It also has very low memory requirements. Its internal round structure benefits from instruction level parallelism. This ability improves its performance.

2.4.1.1 AES Encryption Algorithm

The AES encryption/decryption model is shown in Fig. 2.3, where the AES number of rounds is shown in Fig. 2.3 and it is 10, when the encryption key length is 128 bits long. The number of rounds is 12 with 192 bits key length, and 14 with 256 bits key length. At any round, the system input data is XORed with first four words of the key array ($W_0 - W_3$). At the decryption process, we XOR the decrypted data array with the last four words of the key array. In the AES encryption algorithm, each round consists of four steps as follows [10–12]:

1) Substitute bytes.
2) Shift rows.
3) Mix columns.
4) Add round key.

The last process will be XORing the output data of the four steps described before with four words from the key array [10–12]. The last round for encryption does not involve the "Mix columns" step and the last round for decryption does not involve the "Inverse mix columns" step.

The four steps in each round of processing are shown in Fig. 2.3 and are described as follows:

1. Sub Bytes. It is defined as a forward substitution process, and it is applied in byte-by-byte sequence, where the related substitution process used through the decryption schedule is named Inv Sub Bytes. This step contains a 16 × 16 S-box used to decide the replacement byte of a given byte in the input state array.

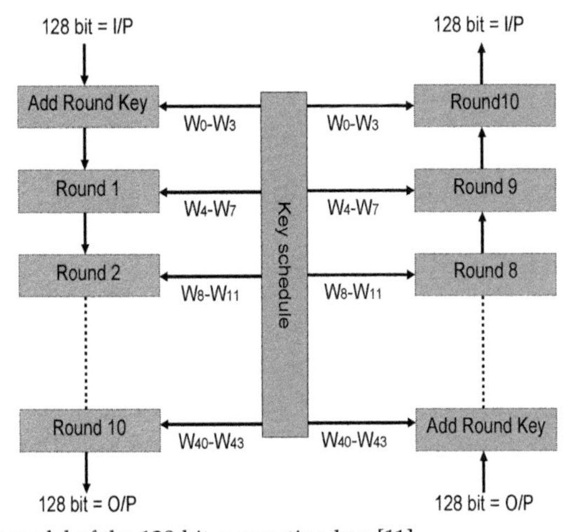

Fig. 2.2: The AES model of the 128-bit encryption key [11].

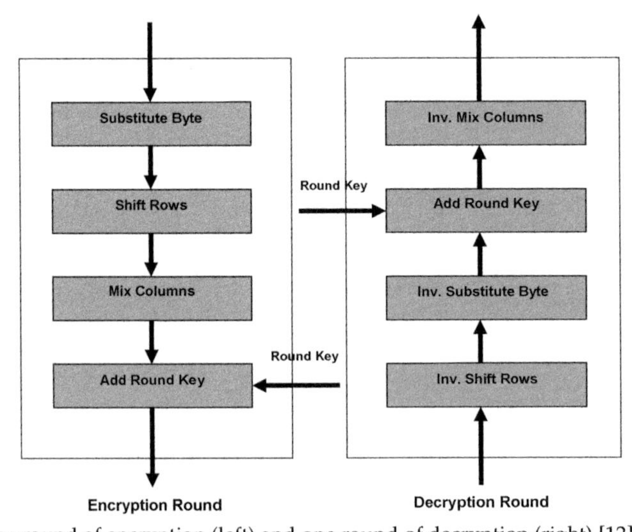

Fig. 2.3: One round of encryption (left) and one round of decryption (right) [12].

2. Shift Rows. It is defined as a forward array shifting of the rows state. Through the decryption schedule, we may call it Inv Shift Rows. For inverse transformation, scrambling is the goal of this process inside each 128-bit block.

3. Mix Columns. It is defined as a forward mixing process of each byte, separately, in each column. Also, the corresponding process through the decryption schedule is called Inv Mix Columns, and it contains an inverse mix column transformation. Our goal here is scrambling of the 128-bit input block.

4. Add Round Key. It is defined as the round key adding to the output data of the previous step. The corresponding stage through the decryption schedule is called Inv Add Round Key.

2.4.2 The RC6 Algorithm

The design of the RC6 algorithm is based on RC5 [11]. Modifications have been performed on RC5 in order to increase security and improve performance. RC6 is one of the 15 candidate algorithms that were presented at the first AES candidate conference in August 1998. It was submitted by Rivest Shamir Adelman (RSA) laboratories [12] and has been selected as one of the five finalists for the second round of the AES contest organized by the NIST [13–14].

2.4.2.1 RC6 Encryption Algorithm

The RC6 algorithm depends on four processing registers, each register works with 32 bits to handle 128 bits as a total size of the input or output block. The RC6 is parameterized by the word size (w) in bits, the number of rounds (r), and the encryption/decryption key length in bytes (b). Figure 2.4 shows the different stages of the RC6 encryption algorithm, where any round in this algorithm has the following five stages:

1. The function described as a squaring one, $f(x) = g(x) \bmod 2^w = x(2x+1) \bmod 2^{32}$, twice.
2. Two modulo 2^{32} additions described in the figure by + in a square.
3. Two XOR operations described by + in a circle.
4. Two fixed right rotations described by <<, and two fixed left rotations, all are data dependent and described in the figure by >>.
5. The XOR operation is used as a mixing operation.

The (f) function, the quantum rotations, and finally the modular additions all give the algorithm its strength, where the number of runs increases the security [9, 15]. The key extends from a b-byte key into a $2r+4$ word array to produce a secret key $S = (S_0, \dots\dots\dots, S_{2r+3})$. The RC6 encryption algorithm is simply performed by using four registers A, B, C and D [16, 17].

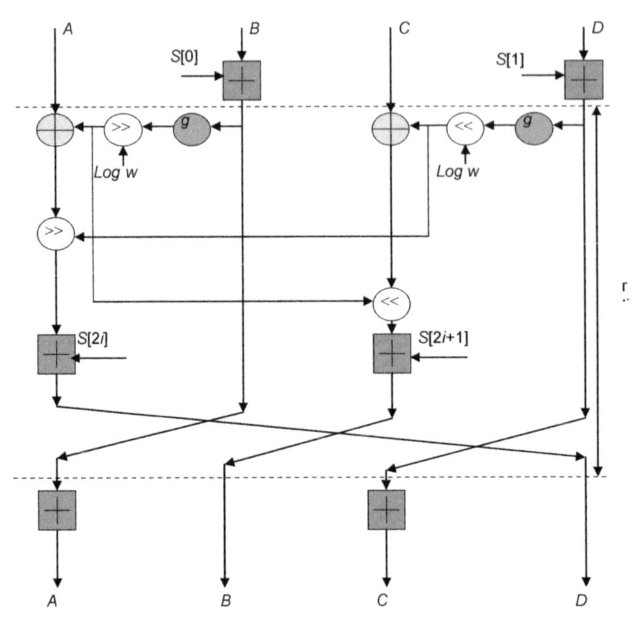

Fig. 2.4: The RC6 encryption algorithm.

2.5 Permutation-based Encryption

Another important branch of image encryption is based on the permutation of pixels to cause confusion. The permutation is simply a pixel rearrangement process. Chaotic maps can achieve this purpose. One of these maps is the Baker map. It is a chaotic map, which converts a unit square into a permutated version of itself.

It assigns a pixel to another pixel in a bijective manner. The discretized Baker map is denoted by $B(v_1,........, v_k)$, where the k is a sequence of integers, where $v_1 + ... + v_k$ is chosen such that each integer v_i divides in to M boxes where $M_i = v_1 + ... + v_i$.

The pixel at indices (l, s) where l and s are the pixel location with $M_i \leq l < M_i + v_i$ and $0 \leq s < M$ is mapped to [18–20]:

$$B_{(n_1,.....,n_k)}(l,s) = \left[\frac{M}{v_i}(l - M_i) + s \bmod \frac{M}{v_i}, \frac{v_i}{M}(s - s \bmod \frac{M}{v_i}) + M_i\right] \quad (2.1)$$

This formula is based on the following steps [17–24]:

1) An $M \times M$ square matrix is divided into k rectangles of width v_i and number of elements M.
2) The elements in each rectangle are rearranged to a row in the permuted rectangle. Rectangles are taken from right to left, beginning with upper rectangles then lower ones.

3) Inside each rectangle, the scan begins from the bottom left corner towards upper elements.

Figure 2.5 shows an example of the chaotic interleaving of an (8 × 8) square matrix (i.e., $M = 8$). The secret key is $S = [n_1, n_2, n_3] = [2, 4, 2]$.

p1	p2	p3	p4	p5	p6	p7	p8
p9	p10	p11	p12	p13	p14	p15	p16
p17	p18	p19	p20	p21	p22	p23	p24
p25	p26	p27	p28	p29	p30	p31	p32
p33	p34	p35	p36	p37	p38	p39	p40
p41	p42	p43	p44	p45	p46	p47	p48
p49	p50	p51	p52	p53	p54	p55	p56
p57	p58	p59	p60	p61	p62	p63	p64

p31	p23	p15	p7	p32	p24	p16	p8
p63	p55	p47	p39	p64	p56	p48	p40
p11	p3	p12	p4	p13	p5	p14	p6
p27	p19	p28	p20	p29	p21	p30	p22
p43	p35	p44	p36	p45	p37	p46	p38
p59	p51	p60	p52	p61	p53	p62	p54
p25	p17	p9	p1	p26	p18	p10	p2
p57	p49	p41	p33	p58	p50	p42	p34

Fig. 2.5: Chaotic encryption of an 8 × 8 matrix with $S = [2, 4, 2]$ [19].

2.5.1 CBC, CFB, and OFB Chaotic Encryption Modes

The general model of a typical encryption system could be described by Eq. (2.2), where P is the plaintext video frame, E is the encryption algorithm, K is the encryption key, and C is the ciphertext video frame, the decryption procedure is given by Eq. (2.3), where D is the decryption algorithm, and K' is the decryption key. It may or may not be the same as the encryption key, K. The multimedia chaotic encryption can be done with three modes of operation, the Cipher Block Chaining (CBC) mode, the Cipher Feed Back (CFB) mode, and the Output Feed Back (OFB) mode [14]. The three modes are used to decide which one of them will increase the security of the transmitted data.

$$E(P, K) = C \tag{2.2}$$

$$D(C, K') = P \tag{2.3}$$

The CBC is a mode of operation for a block cipher in which a sequence of bits is encrypted as a single unit or block with a cipher key applied to the entire block. Cipher block chaining uses an IV of a certain length. One of its key characteristics is that it uses a chaining mechanism that causes the decryption of a block cipher of the ciphertext to depend on all the preceding ciphertext blocks [21, 22]. In the CBC mode, each block of the plaintext is XORed with the previous ciphertext block before being encrypted. So, each ciphertext block is dependent on all plaintext blocks up to that point. In decryption, the same XOR operation is repeated so that its effect is cancelled. This mechanism is shown in Fig. 2.6.

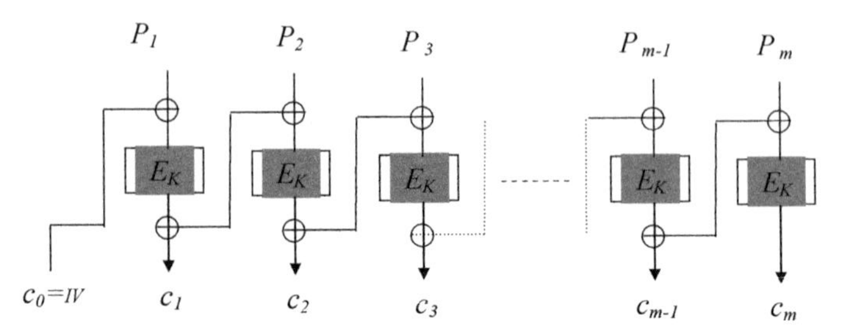

Fig. 2.6: The encryption process using the CBC mode.

The main disadvantage of the CBC mode is that an error in one ciphertext block impacts two plaintext blocks upon decryption. In the CBC mode, the encryption algorithm is given by Eq. (2.3) and the decryption algorithm is given by Eq. (2.4), where $j = 1, 2, 3\ldots\ldots$ and $C_0 = \text{IV}$.

$$C_j = E_k (C_{j-1} \oplus P_j) \tag{2.4}$$

$$P_j = D_k (C_j) \oplus C_{j-1} \tag{2.5}$$

The CFB is an encryption mode in which the plaintext block is XORed with the output of the previous stage performed on the preceding block. Like CBC mode, the CFB mode makes use of an IV [21, 23]. The XOR operation hides the plaintext patterns. The plaintext cannot be directly worked on unless there is retrieval of blocks from either the beginning or end of the ciphertext. Figure 2.7 illustrates the CFB mode. In the CFB mode, the encryption algorithm is given by Eq. (2.5) and the decryption algorithm is given by Eq. (2.6), where $I_j = E_k (C_{j-1})$, $j = 1, 2, 3\ldots\ldots$, and $C_0 = \text{IV}$.

$$C_j = P_j \oplus I_j \tag{2.6}$$

$$P_j = C_j \oplus I_j \tag{2.7}$$

The OFB mode has some similarities with the CFB mode, in that it permits encryption of different block sizes but has the main difference that each plaintext block is XORed directly with the encryption result of the preceding block. It also uses an IV. Changing the IV for the same plaintext block results in different ciphertexts [18, 19]. The OFB mode is illustrated in Fig. 2.8. In this mode, the encryption algorithm is given by Eq. (2.7) and the decryption algorithm is given by Eq. (2.8), where $I_j = E_k (I_{j-1})$, $j = 1, 2, 3\ldots\ldots$, and $C_0 = \text{IV}$.

$$C_j = P_j \oplus I_j \tag{2.8}$$

$$P_j = C_j \oplus I_j \tag{2.9}$$

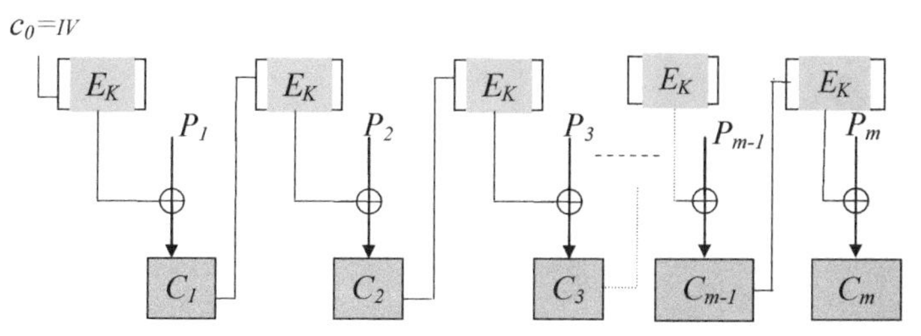

Fig. 2.7: The encryption process using the CFB mode.

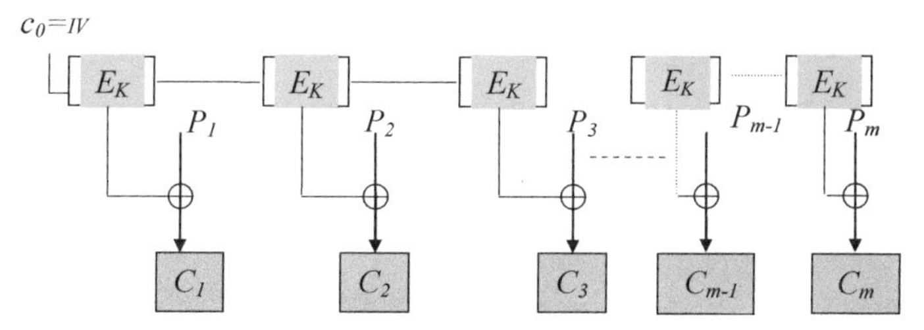

Fig. 2.8: The encryption process using the OFB mode.

2.6 Encryption Quality Metrics

We need to use some mathematical metrics to evaluate the degree of encryption quality. We will use the histogram uniformity and the deviation from the original image.

2.6.1 Histogram

The histogram of an image represents the occurrence probability of each gray level in the image. It is required to be as uniform as possible to achieve security. Figure 2.9 shows the histograms of the Cameraman images encrypted with different algorithms. It is clear from this figure that histogram uniformity is achieved with diffusion-based algorithms. The histogram of a chaotic encrypted image is the same as that of the original image, because chaotic encryption is based on permutations only.

2.6.2 Deviation of an Encrypted Image

The deviation between the original image and the encrypted image is expressed as follows:

$$D = \sum_{i=0}^{g \times g} |I_i - J_i| \tag{2.10}$$

where *Ii* is the ith pixel of the original image and *Ji* is the *J*th pixel of the encrypted image [13], [14], [24], and [29].

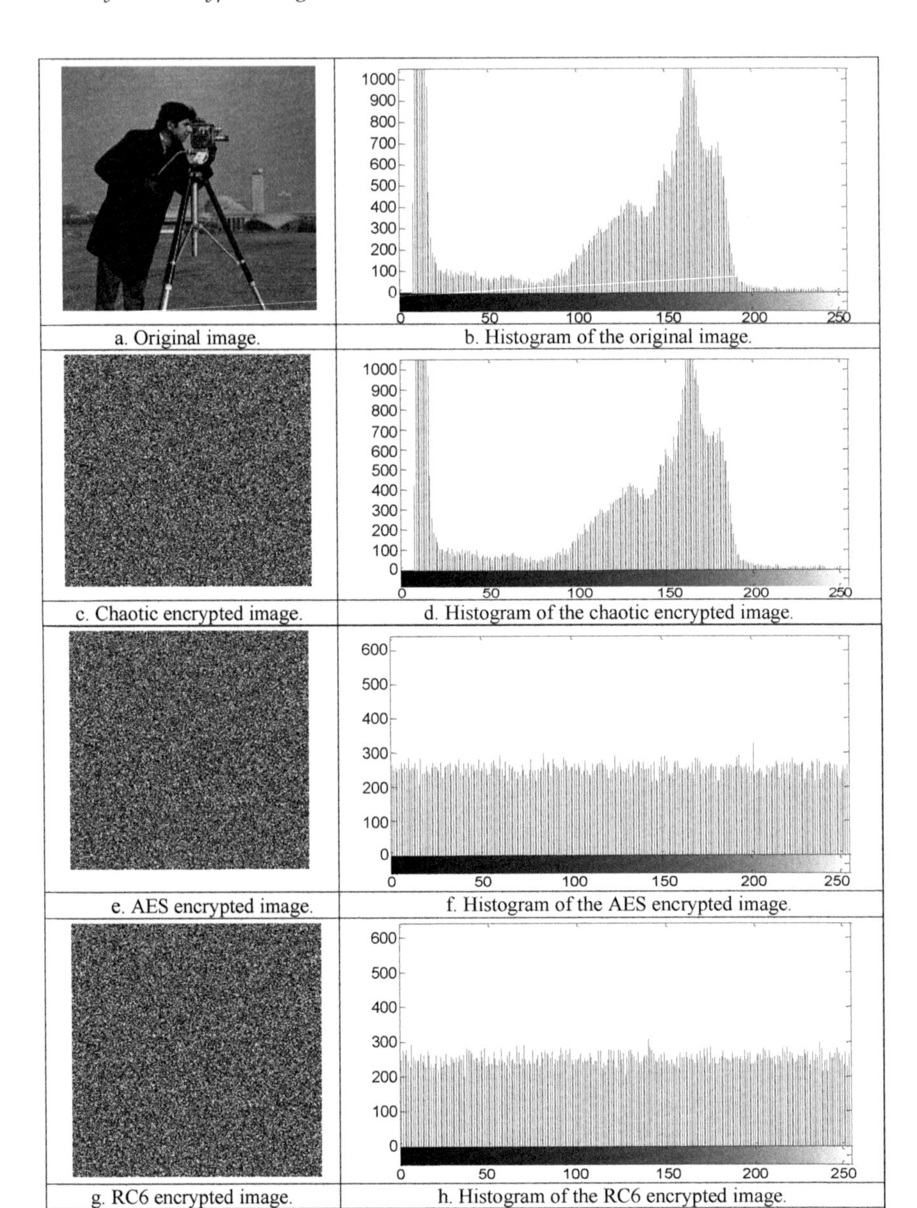

a. Original image. b. Histogram of the original image.
c. Chaotic encrypted image. d. Histogram of the chaotic encrypted image.
e. AES encrypted image. f. Histogram of the AES encrypted image.
g. RC6 encrypted image. h. Histogram of the RC6 encrypted image.

Fig. 2.9: Original and encrypted images with their histograms.

Table 2.1 shows the deviation values for the Cameraman images encrypted with different algorithms. It is clear from this table that diffusion-based algorithms are preferred from the deviation perspective due to their diffusion property, which enhances the security. So, as the deviation value is increased, the degree of security is enhanced.

Table 2.1: Deviation values for the encrypted Cameraman images.

Encryption Technique	Chaotic Encryption	DES Encryption	AES Encryption	RC6 Encryption
D	33.73	35.39	35.17	35.11

2.6.3 Immunity to Noise

Noise immunity reflects the ability of the image cryptosystem to tolerate noise. To test noise immunity, error patterns with different Bit Error Rates (BERs) are simulated and added to the encrypted images, and then the decryption algorithm is performed. If the decrypted image is close to the original one, we can say that the cryptosystem at hand is immune to noise. As the PSNR of the decrypted image is increased, the algorithm becomes more immune to noise. Figure 2.10 shows the variation of the PSNR of the decrypted image with the simulated BER. From this figure, it is clear that chaotic encryption achieves the best PSNR values in the presence of errors.

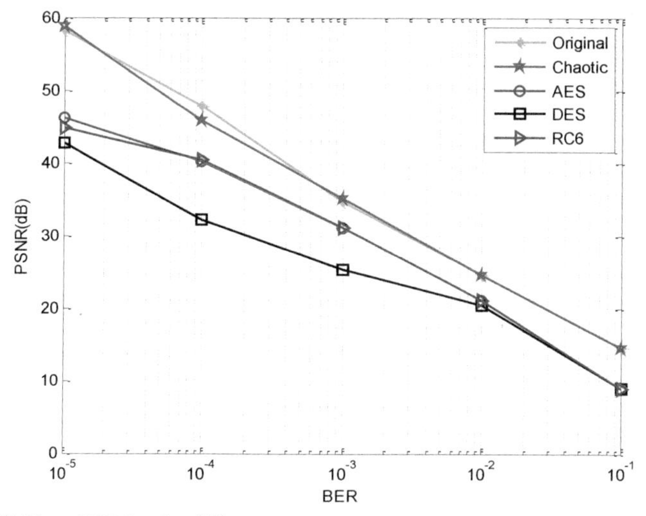

Fig. 2.10: PSNR vs. BER for the different cryptosystems.

Chapter 3

Rubik's Cube Encryption for Wireless Communications

3.1 Overview of the Proposed Hybrid Encryption Approach

In this chapter, the suggested hybrid encryption framework begins with chaotic Baker map permutation, AES, or RC6 algorithm as a first step for encrypting the multiple images, separately. Then, the resulting encrypted images are further encrypted in a second stage with Rubik's cube algorithm. Chaotic, RC6 or AES encrypted images are utilized as the faces of the Rubik's cube. From the concepts of image encryption, the RC6 or AES algorithm introduces a degree of diffusion, whilst the chaotic Baker map adds a degree of permutation.

Moreover, the Rubik's cube algorithm adds more permutation to the encrypted images together. The encrypted images are further transmitted through a wireless channel and decrypted at the receiver side. The quality evaluation of the decrypted images at the receiver side reveals good performance. Simulation results reveal that the suggested hybrid encryption framework is efficient, and it presents strong algorithm against attacks.

In the proposed communication system, the images to be transmitted are first encrypted as illustrated in Fig. 3.1 and Fig. 3.2. The encryption and communication processes can be summarized as follows:

Input the database of images of the database to encrypt;

- initialize all process;
- receive the image bit-streams;
- Start with chaotic or AES, or RC6 encryption algorithm;

for all Encrypted images, ***do***
- Set the encrypted images over Rubik's cube nine faces;
- Convert the whole faces into a 2-D image with larger size;
- Randomize the new image with Rubik's cube mechanism;
- Transform the image into binary format;

end for
if Proposed Hybrid Encryption algorithm is ***done, then***
- Perform OFDM modulation;
- Transmit the signal over the wireless channel;
- Perform channel equalization at the receiver;
- Transform the received signal into binary format;
- Reconstruct the image into pixel values;

end if
for all Received encrypted images, ***do***
- Return to the canonical basis to get the decrypted images;

end for

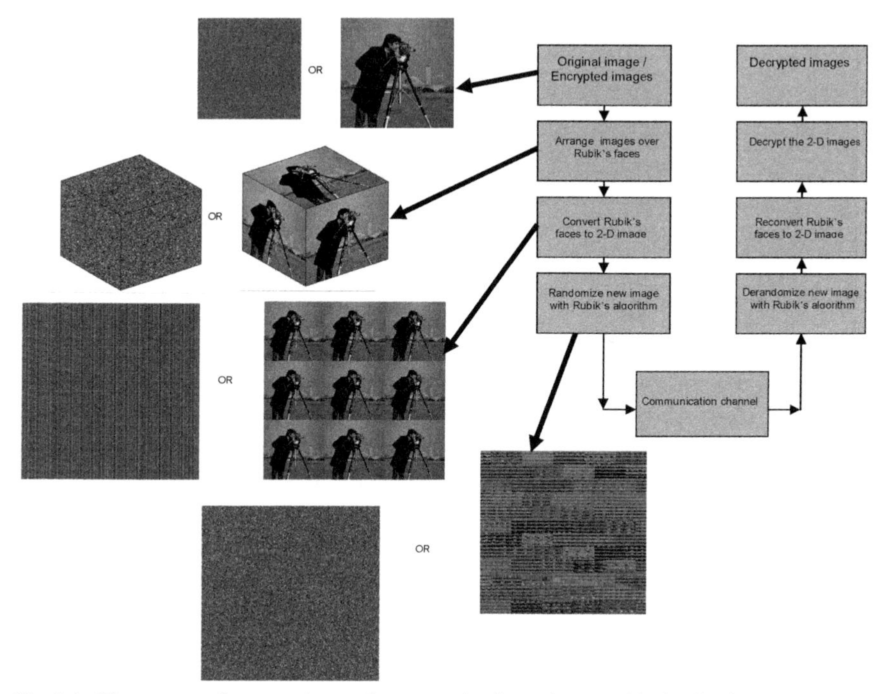

Fig. 3.1: The proposed encryption and communication scheme with similar faces.

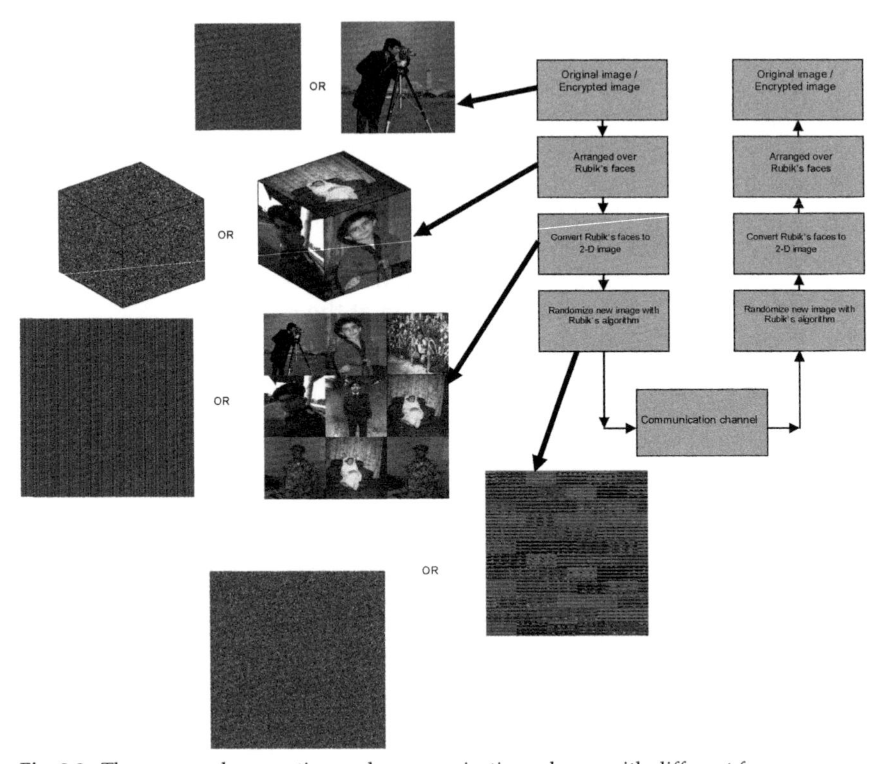

Fig. 3.2: The proposed encryption and communication scheme with different faces.

3.2 Puzzle Concept

The Rubik's cube proposed in this book depends on the puzzle concept. A puzzle is a problem that tests the ability of the mind imagination, skill, and cleverness of the solver. Puzzles were and are still classified as a form of games and entertainment, but we can also consider them in mathematical problems or logical quizzes in different cases. People with a high attitude may be better than others, when they are dealing with those puzzles. Also, those quizzes may be solved easily by computer programs [62]. Generally, there are four different categories of puzzles. Some are listed as follows [63].

1. **Logical puzzles.** These are problems defined as recreational mathematics, such as knight's tour on a chess board.
2. **Mathematical puzzles.** In these puzzles, we find a missing number in a chain, a missing number in a table, a missing time, or a missing distance.
3. **Paper and pencil puzzles.** In these puzzles, we combine the points to build an image using a maze puzzle, for example.

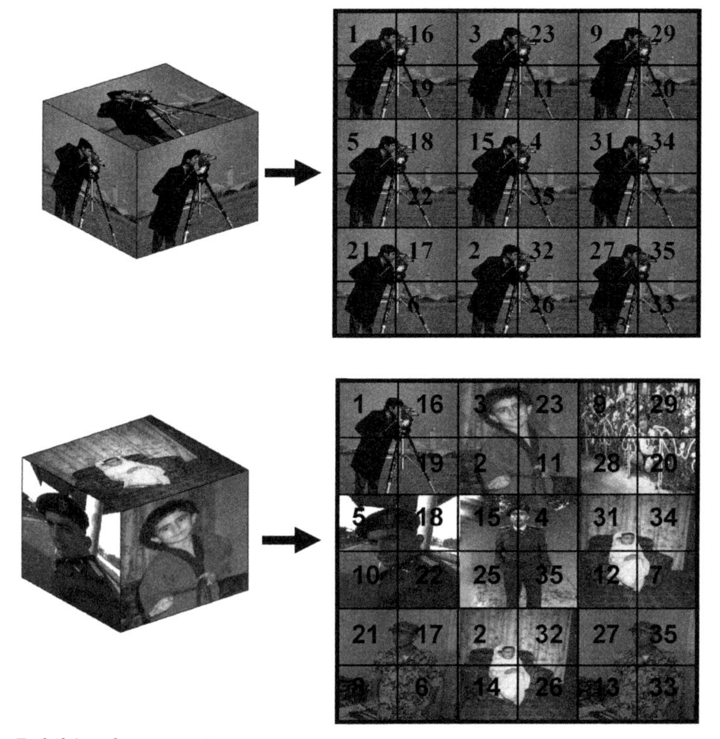

Fig. 3.3: Rubik's cube concept.

4. **Mechanical puzzles.** These puzzles include the burr puzzle, soma cube, and Rubik's cube, as shown in Fig. 3.3.

The 2-D Rubik's cube in its discretized version can be used for this purpose. Let $B(n_1, ..., \underline{n}_k)$ denote the discretized image, where the vector $[n_1, ..., \underline{n}_k]$ represents the secret key, in which n is the number of columns in map S_{key}. Defining N as the number of data items in one row, the secret key is chosen such that each integer n_i divides I, $n_1 + ... + \underline{n}_k = I$. Let $I_i = n_1 + ... + n_i$, $I_i \leq r < I_i + n_i$ and $0 \leq s < I$ where (r, s) are the pixel location in map [64].

$$B_{(n_1,.....,n_k)}(r,s) = \left[\frac{I}{n_i}(r - I_i) + s \bmod \frac{I}{n_i}, \frac{n}{I}\left(s - s \bmod \frac{I}{n_i}\right) + I_i\right] \tag{3.1}$$

In steps, the Rubik's permutation is performed as follows [65, 66]:

1. An $I \times I$ square matrix is divided into k rectangles of width n_i and number of elements I.
2. The elements in each rectangle are rearranged to a row in the permuted rectangle. Rectangles are taken from left to right, beginning with upper

Fig. 3.4: The Cameraman image and the encrypted version.

rectangles then lower ones. Inside each rectangle, the scan begins from the bottom left corner towards upper elements [66, 67]. Figure 3.4 depicts the Cameraman image and its encrypted version.

3.3 AES with Rubik's Cube Algorithm

The AES algorithm is a block cipher algorithm that uses an encryption key and several encryption rounds. In our study of the AES, the block is 128 bits or 16 bytes in length. The term "rounds" refers to the way in which the encryption algorithm mixes the data re-encryption depending on the length of the key. Rubik's cube algorithm depends on the puzzle principle that randomizes the data. Mixing the two algorithms will develop both the transmission quality and the encryption strength. With advanced and powerful computers, the DES standard has proved to be insecure.

As a result, in 1997, the NIST called for proposals for the next generation encryption standard. After three years' work, the NIST announced its selection for the AES algorithm. In 2001, the AES has become the official encryption standard. It is a block-structured algorithm with variable-length keys of 128 bits, 192 bits, and 256 bits. This, in itself, significantly increases the security level compared to the DES [68–69]. The algorithm has very good performance in both hardware and software implementations. It also has very low memory requirements. This property improves its performance. On the other hand, the Rubik's cube encryption algorithm was invented by the Hungarian architect Erno Rubik in 1974 as an advanced puzzle game and was then called "The Magic Cube". It was renamed after its inventor in 1980.

In the past few years, several encryption algorithms based on chaotic systems have been proposed for digital image protection against cryptographic attacks. These encryption algorithms typically use relatively small key spaces, and thus offer limited security. So, we propose a novel image encryption algorithm based on Rubik's cube principle [70]. In this algorithm, the original image is scrambled using the AES algorithm, and then it is randomized by Rubik's cube algorithm. The simulation results and security analysis show that the proposed image encryption scheme not only achieves a good encryption performance, but also resists attacks and guarantees high transmission quality, especially at large *Eb/N0* [70, 71].

3.3.1 Histogram Analysis

The proposed algorithm gives approximate flat histograms, which means that the AES + Rubik's cube algorithm achieves perfect hiding of images. For a fair study, we apply simulation programs to the two images shown in Fig. 3.5 and Fig. 3.6, where the first is a built-in image and the second is taken through a mobile-camera.

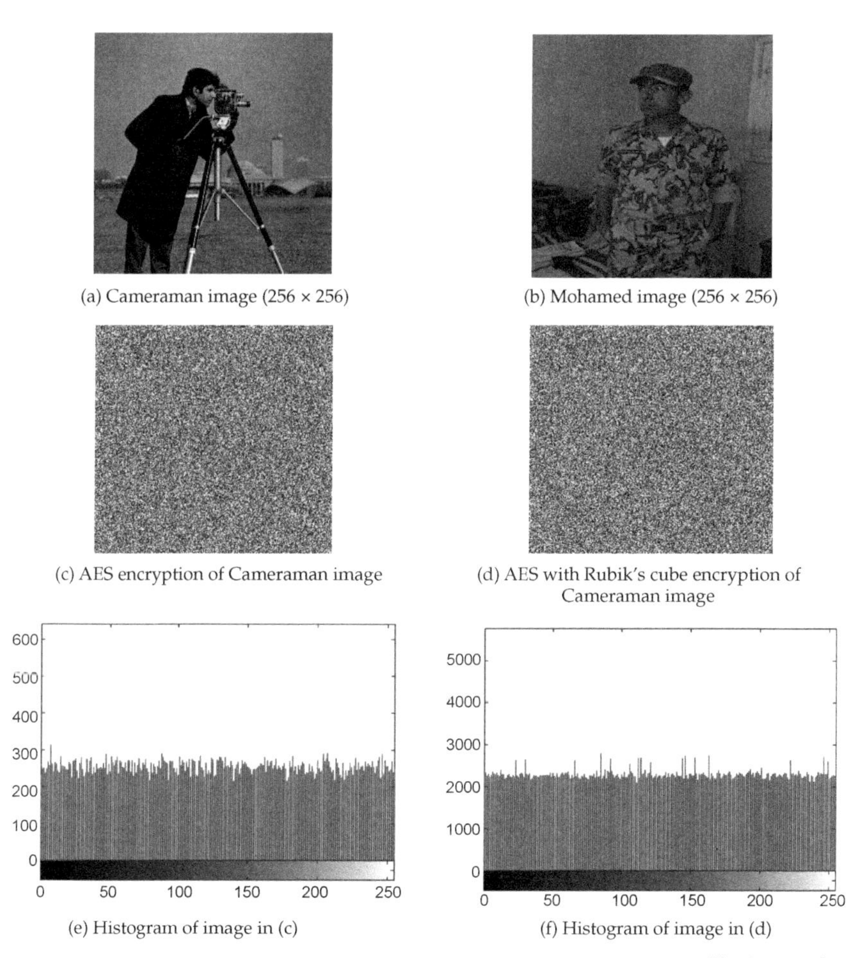

(a) Cameraman image (256 × 256)

(b) Mohamed image (256 × 256)

(c) AES encryption of Cameraman image

(d) AES with Rubik's cube encryption of Cameraman image

(e) Histogram of image in (c)

(f) Histogram of image in (d)

Fig. 3.5 contd. ...

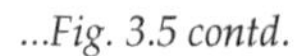

...Fig. 3.5 contd.

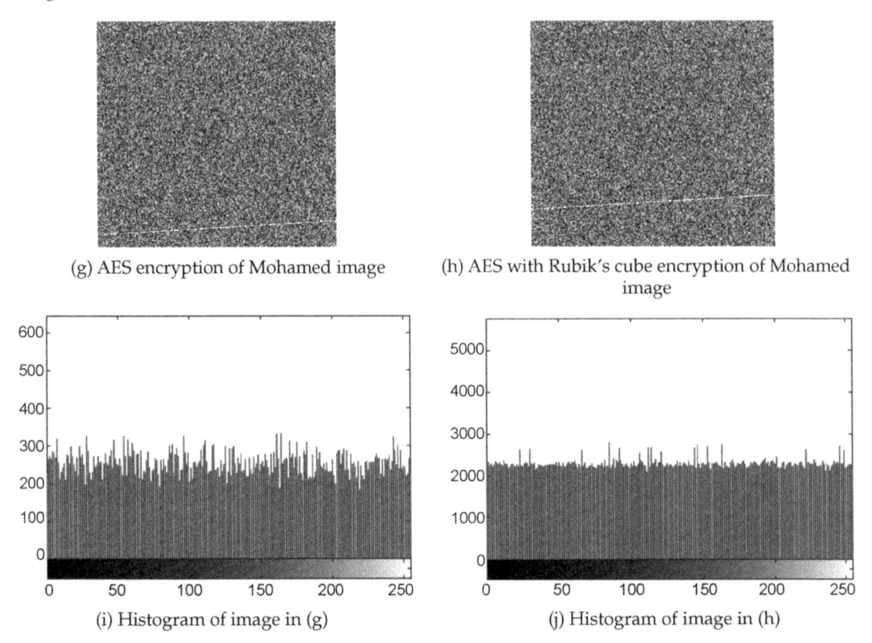

(g) AES encryption of Mohamed image

(h) AES with Rubik's cube encryption of Mohamed image

(i) Histogram of image in (g)

(j) Histogram of image in (h)

Fig. 3.5: Encrypted image and histogram of "AES and Rubik's cube" algorithm vs. "AES" algorithms for Cameraman and Mohamed images for the same image on all faces.

(a) Cameraman image (256 × 256)

(b) Mohamed image (256 × 256)

(c) AES encryption of Cameraman image

(d) AES with Rubik's cube encryption of Cameraman image

Fig. 3.6 contd. ...

...*Fig. 3.6 contd.*

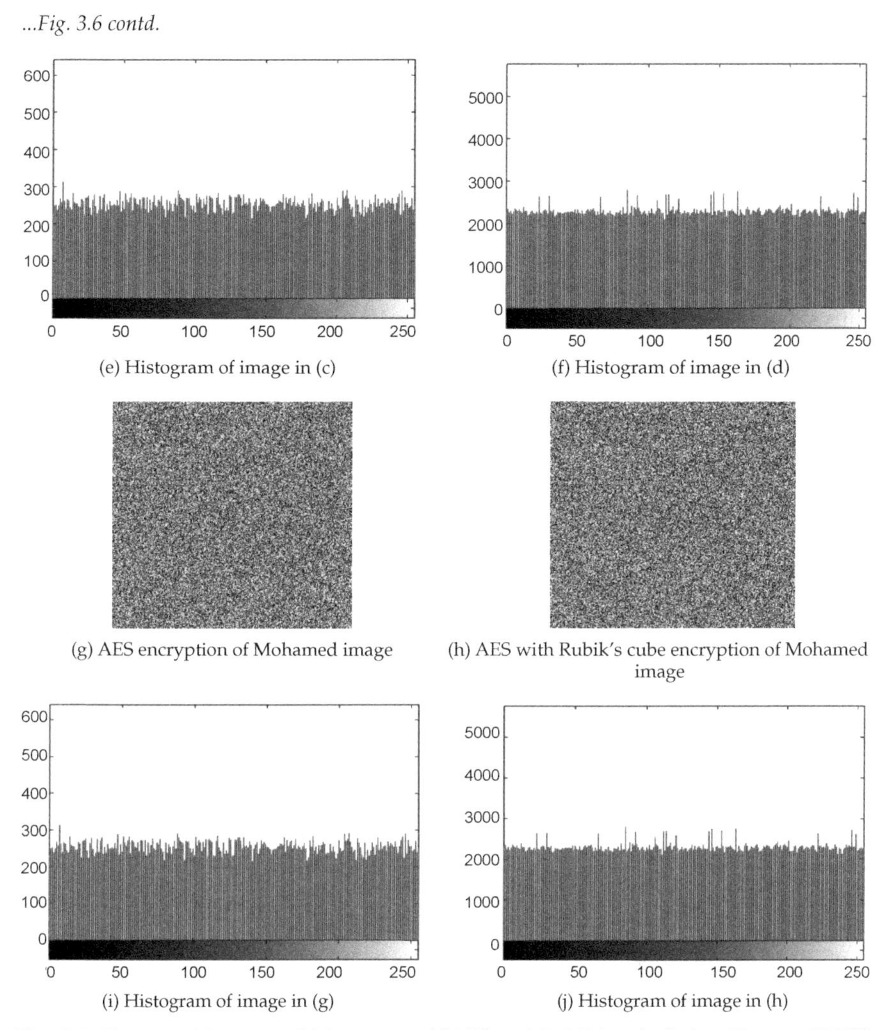

(e) Histogram of image in (c)

(f) Histogram of image in (d)

(g) AES encryption of Mohamed image

(h) AES with Rubik's cube encryption of Mohamed image

(i) Histogram of image in (g)

(j) Histogram of image in (h)

Fig. 3.6: Encrypted image and histogram of "AES and Rubik's cube" algorithm vs. "AES" algorithms for Cameraman and Mohamed images for different image on all faces.

3.3.2 Deviation

The deviation results obtained with AES + Rubik's encryption of Cameraman and Mohamed images are better than those obtained with AES only. The deviation increases as the algorithm becomes stronger, as tabulated in Table 3.1.

3.3.3 Correlation Coefficient

The smallest correlation values are obtained for AES + Rubik's cube encryption. The correlation coefficient decreases as the algorithm becomes strong, as shown in Table 3.2.

3.3.4 Processing Time

The processing times of different encryption algorithms are tabulated in Table 3.3. The recorded times reveal that the AES + Rubik's cube encryption with similar faces record the lowest computational times.

Table 3.1: Deviation values for the encrypted images with different algorithms.

Encryption Algorithm	Deviation
AES encryption of Cameraman	35.18
ASE + Rubik's cube encryption of Cameraman (Similar faces)	35.42
ASE + Rubik's cube encryption of Cameraman (Different faces)	35.25
AES encryption of Mohamed	29.73
ASE + Rubik's cube encryption of Mohamed (Similar faces)	29.85
ASE + Rubik's cube encryption of Mohamed (Different faces)	35.36

Table 3.2: Correlation coefficient values between encrypted and original images for different encryption algorithms.

Encryption Algorithm	Correlation Coefficient
AES encryption of Cameraman	0.0077
ASE + Rubik's cube encryption of Cameraman (Similar faces)	–0.0069
ASE + Rubik's cube encryption of Cameraman (Different faces)	–0.0034
AES encryption of Mohamed	–4.2268 e-04
ASE + Rubik's cube encryption of Mohamed (Similar faces)	–0.0059
ASE + Rubik's cube encryption of Mohamed (Different faces)	–0.0067

Table 3.3: Processing times for different encryption algorithms.

Encryption Algorithm	Processing Time
AES encryption of Cameraman	122.57 seconds
ASE + Rubik's cube encryption of Cameraman (Similar faces)	130.83 seconds
ASE + Rubik's cube encryption of Cameraman (Different faces)	725.17 seconds
AES encryption of Mohamed	112.98 seconds
ASE + Rubik's cube encryption of Mohamed (Similar faces)	117.64 seconds
ASE + Rubik's cube encryption of Mohamed (Different faces)	830.07 seconds

3.3.5 The Noise Immunity

The sensitivity of the encryption algorithms to the channel noise is illustrated in Fig. 3.7, in which the variation of the PSNR of the reconstructed decrypted image with the BER resulting from channel effects is given. It is found that the AES algorithm is more immune to channel disturbance and noise than the AES + Rubik's algorithm.

The sensitivity of the AES encryption and the AES + Rubik's cube encryption algorithms to the BER due to the channel effect is well illustrated in Fig. 3.8. By experimentation, it is found that the AES algorithm is more immune to channel disturbance and noise than the AES + Rubik's cube algorithm.

3.4 RC6 with Rubik's Cube Algorithm

The proposed cryptosystem is based on Rubik's cube method mixed with chaotic Baker map. The IV of the chaotic map works as the main key, and the block size W is a square lattice of $N \times N$ pixels [4–5].

The Rubik's cube algorithm that will be adopted in this book in order to achieve the permutation in the encryption process has nine faces. It will be

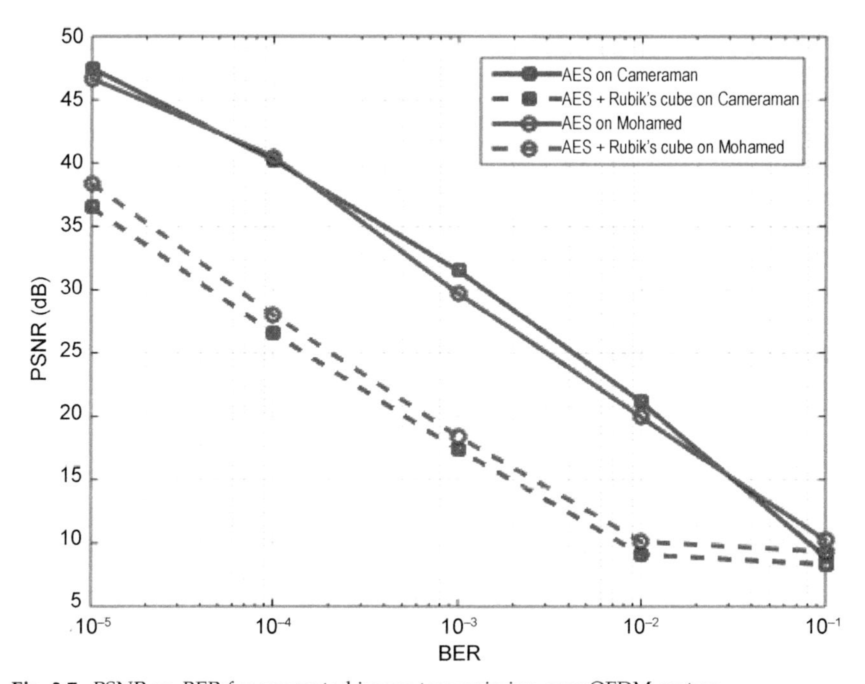

Fig. 3.7: PSNR vs. BER for encrypted image transmission over OFDM system.

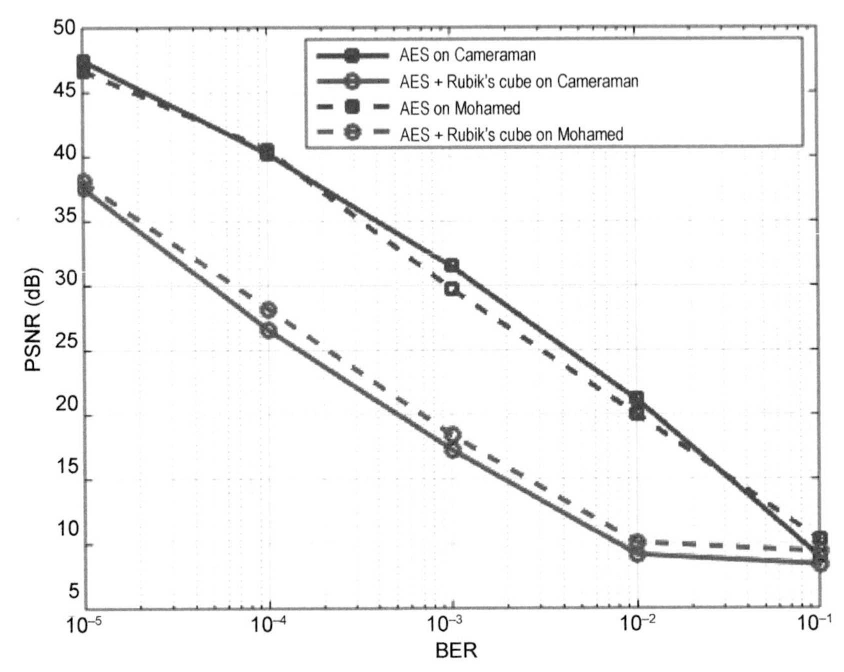

Fig. 3.8: PSNR vs. BER for "AES and Rubik's cube" encrypted image transmission over OFDM system with different faces for the Rubik's cube.

used to encrypt a group of images, simultaneously. The images implemented on these nine faces are those encrypted firstly with the RC6 algorithm. The proposed two-step encryption algorithm will guarantee both diffusion and permutation in the encrypted images [72, 73]. We will study its sensitivity to the wireless channel impairments, and study also the effect of channel equalization on the received image quality.

3.4.1 Histogram Analysis

The RC6 with Rubik's cube algorithm gives near-flat histograms, which implies a high quality of encryption. We have applied the simulation study on the three images shown in Figs. 3.9 and 3.10.

The Rubik's cube key is the IVs for each image and the RC6 key is $(00010203040506070809)_{16}$. The obtained results show good quality of encryption.

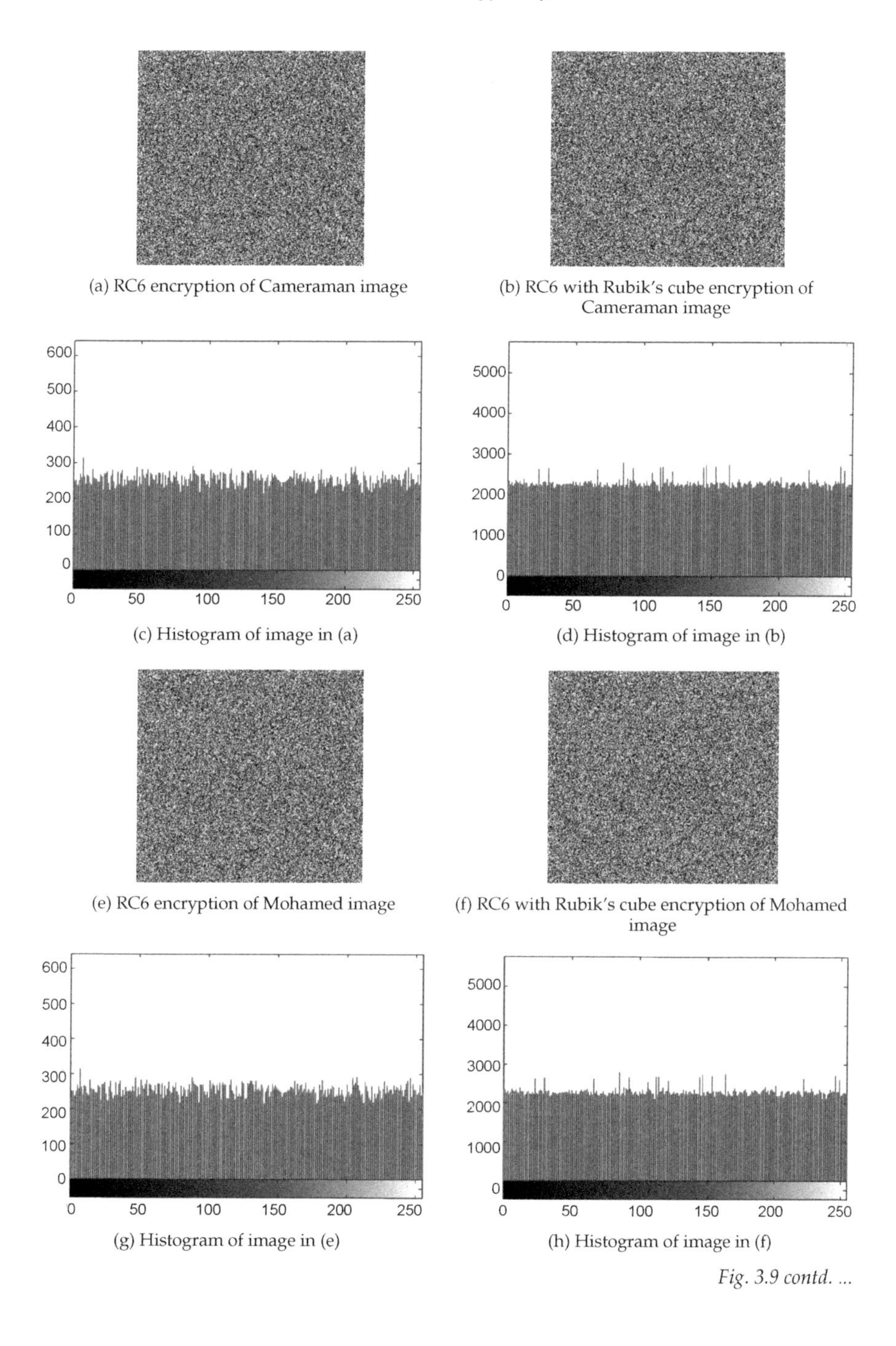

(a) RC6 encryption of Cameraman image

(b) RC6 with Rubik's cube encryption of Cameraman image

(c) Histogram of image in (a)

(d) Histogram of image in (b)

(e) RC6 encryption of Mohamed image

(f) RC6 with Rubik's cube encryption of Mohamed image

(g) Histogram of image in (e)

(h) Histogram of image in (f)

Fig. 3.9 contd. ...

...Fig. 3.9 contd.

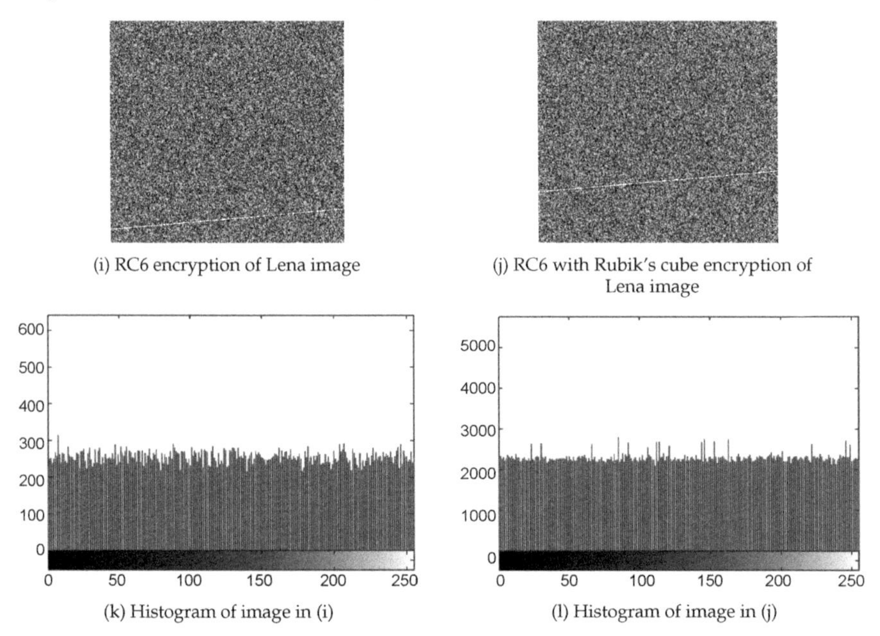

(i) RC6 encryption of Lena image

(j) RC6 with Rubik's cube encryption of Lena image

(k) Histogram of image in (i)

(l) Histogram of image in (j)

Fig. 3.9: Encrypted image and histogram of "RC6 and Rubik's cube" algorithm vs. "RC6" algorithms for Cameraman, Mohamed and Lena images for the same image on all faces.

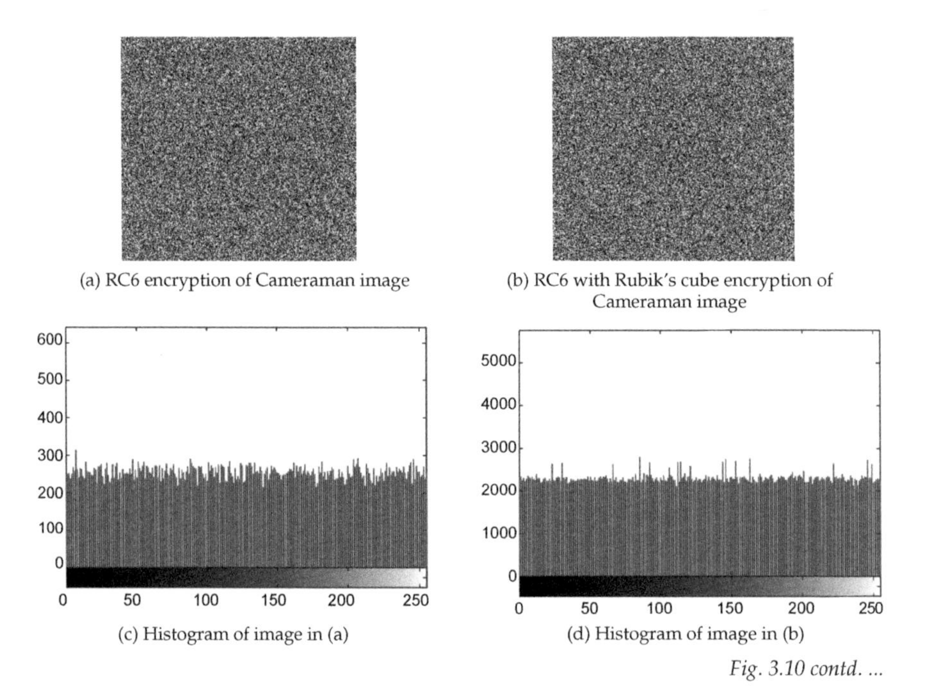

(a) RC6 encryption of Cameraman image

(b) RC6 with Rubik's cube encryption of Cameraman image

(c) Histogram of image in (a)

(d) Histogram of image in (b)

Fig. 3.10 contd. ...

...Fig. 3.10 contd.

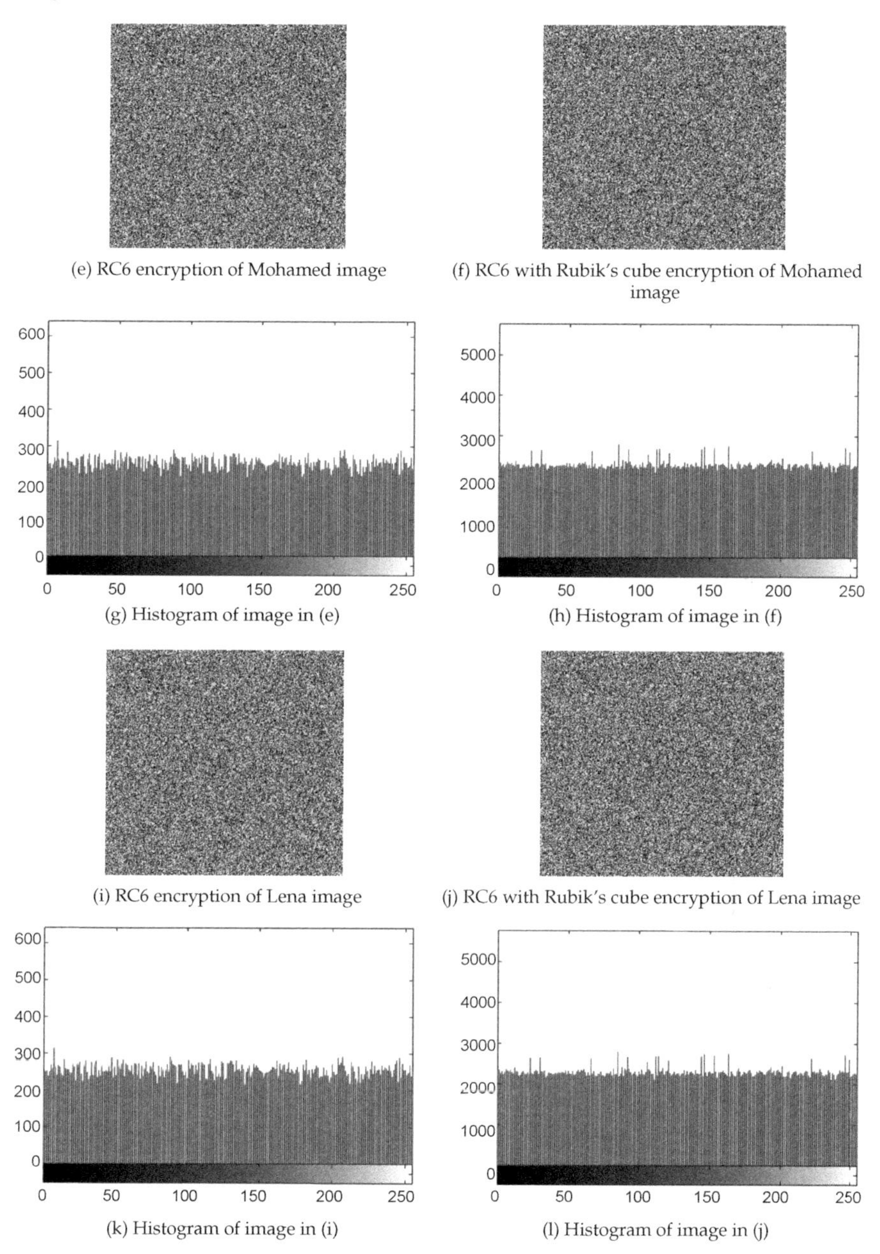

(e) RC6 encryption of Mohamed image

(f) RC6 with Rubik's cube encryption of Mohamed image

(g) Histogram of image in (e)

(h) Histogram of image in (f)

(i) RC6 encryption of Lena image

(j) RC6 with Rubik's cube encryption of Lena image

(k) Histogram of image in (i)

(l) Histogram of image in (j)

Fig. 3.10: Encrypted image and histogram of "RC6 and Rubik's cube" algorithm vs. "RC6" algorithms for Cameraman, Mohamed and Lena images for different images on all faces.

3.4.2 Deviation

The deviation of RC6 with Rubik's cube method is better than that without Rubik's cube method on Cameraman, Mohamed, and Lena images.

Table 3.4: Deviation results for different methods.

Encryption Algorithms	Deviation
RC6 encryption of Cameraman	35.11
RC6 + Rubik's cube encryption of Cameraman (Similar faces)	35.35
RC6 + Rubik's cube encryption of Cameraman (Different faces)	35.61
RC6 encryption of Mohamed	29.66
RC6 + Rubik's cube encryption of Mohamed (Similar faces)	29.71
RC6 + Rubik's cube encryption of Mohamed (Different faces)	30.00
RC6 encryption of Lena	34.39
RC6 + Rubik's cube encryption of Lena (Similar faces)	34.54
RC6 + Rubik's cube encryption of Lena (Different faces)	34.66

3.4.3 Correlation Coefficient

The smallest correlation is for the algorithm containing RC6, and Rubik's cube method on Cameraman image.

Table 3.5: Correlation coefficient results for different methods.

Encryption Algorithm	Correlation
RC6 encryption of Cameraman	0.0023
RC6 + Rubik's cube encryption of Cameraman (Similar faces)	-3.3671*10e-04
RC6 + Rubik's cube encryption of Cameraman (Different faces)	-0.0016
RC6 encryption of Mohamed	0.0042
RC6 + Rubik's cube encryption of Mohamed (Similar faces)	0.0060
RC6 + Rubik's cube encryption of Mohamed (Different faces)	0.0026
RC6 encryption of Lena	-8.6183e-04
RC6 + Rubik's cube encryption of Lena (Similar faces)	-0.0044
RC6 + Rubik's cube encryption of Lena (Different faces)	-0.0043

3.4.4 Processing Time

The calculations reveal that the RC6 + Rubik's cube algorithm needs less time in processing as shown in Table 3.6.

Table 3.6: Processing time results for different algorithms.

Encryption Algorithm	**Processing Time**
RC6 encryption of Cameraman	653.20 seconds
RC6 + Rubik's cube encryption of Cameraman (Similar faces)	681.66 seconds
RC6 + Rubik's cube encryption of Cameraman (Different faces)	9806.57 seconds
RC6 encryption of Mohamed	724.41 seconds
RC6 + Rubik's cube encryption of Mohamed (Similar faces)	729.72 seconds
RC6 + Rubik's cube encryption of Mohamed (Different faces)	5567.08 seconds
RC6 encryption of Lena	1243.65 seconds
RC6 + Rubik's cube encryption of Lena (Similar faces)	9714.62 seconds
RC6 + Rubik's cube encryption of Lena (Different faces)	9909.56 seconds

3.4.5 Noise Immunity

The sensitivity of RC6 and RC6 + Rubik's algorithms to the BER is revealed in Fig. 3.11. It is clear that the RC6 algorithm is more immune to channel disturbance and noise than the RC6 + Rubik's algorithm, while at large values of the BER, all algorithms for the three different images have approximately the same behavior.

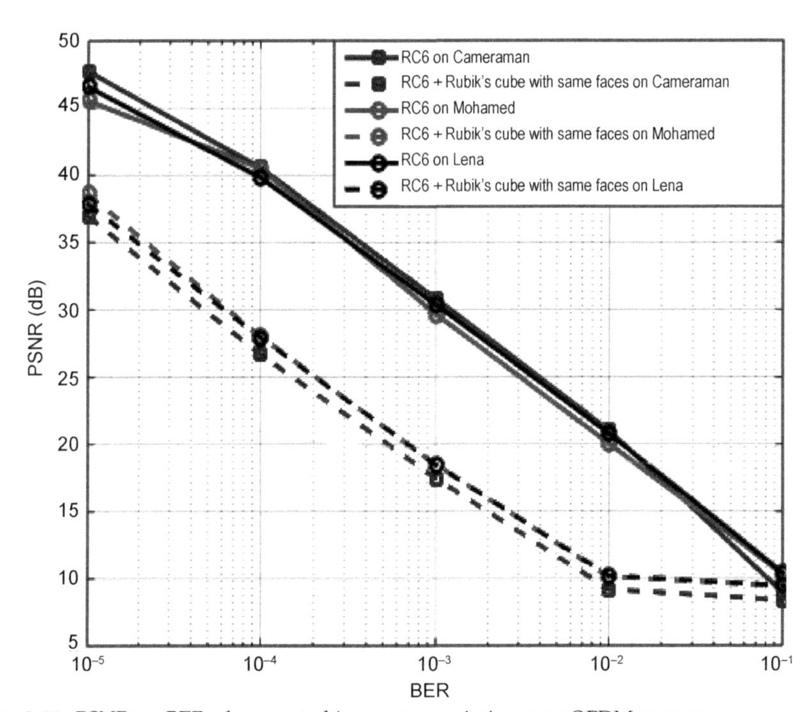

Fig. 3.11: PSNR vs. BER of encrypted image transmission over OFDM system.

3.5 Chaotic Encryption Algorithm in Different Modes with Rubik's Cube Algorithm

3.5.1 Histogram Analysis

For a fair study, we apply the simulation program to two images: Cameraman and Mohamed.

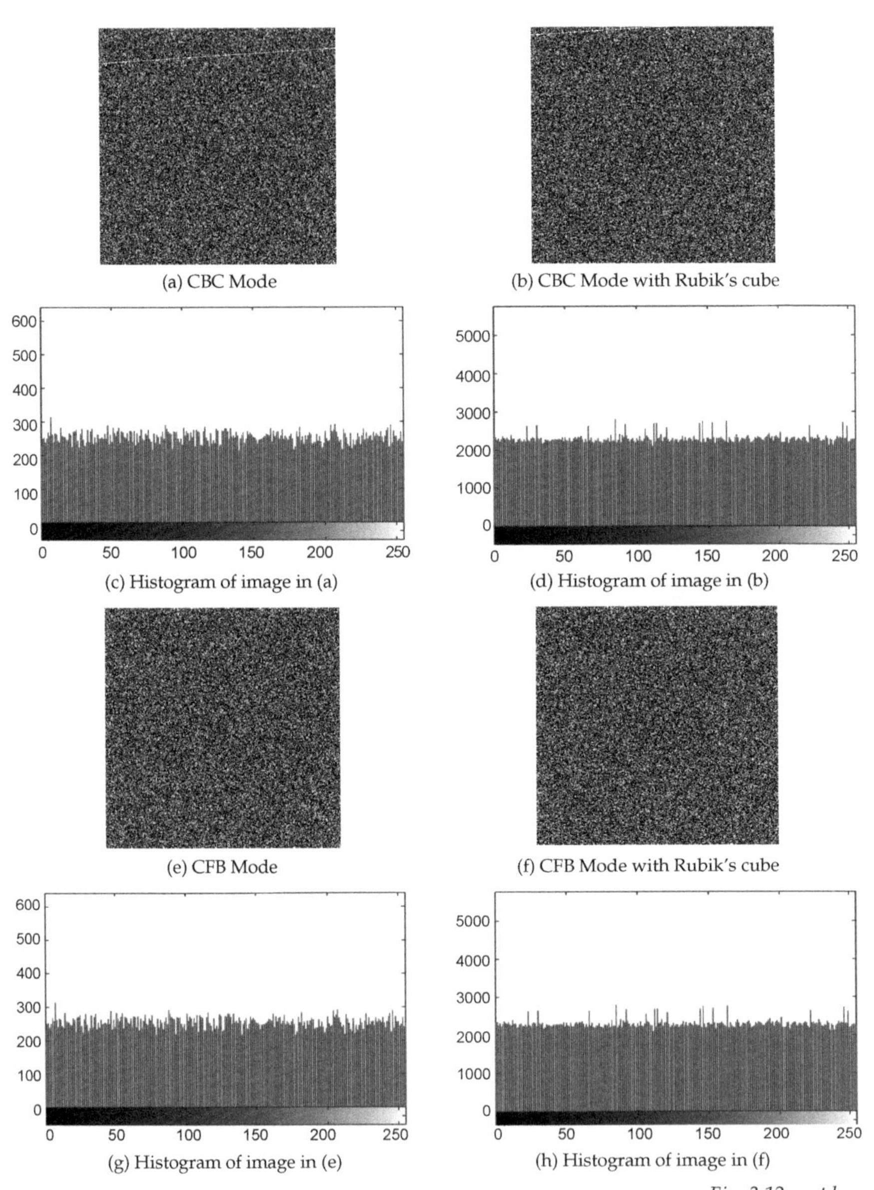

(a) CBC Mode (b) CBC Mode with Rubik's cube

(c) Histogram of image in (a) (d) Histogram of image in (b)

(e) CFB Mode (f) CFB Mode with Rubik's cube

(g) Histogram of image in (e) (h) Histogram of image in (f)

Fig. 3.12 contd. ...

...Fig. 3.12 contd.

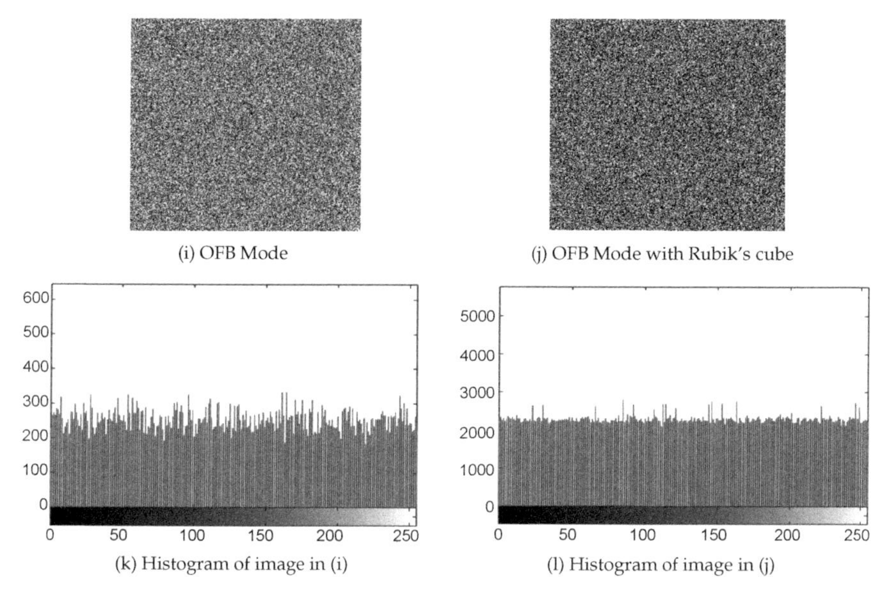

(i) OFB Mode

(j) OFB Mode with Rubik's cube

(k) Histogram of image in (i)

(l) Histogram of image in (j)

Fig. 3.12: Encrypted image and histogram of "Chaotic and Rubik's cube" algorithm output vs. "Chaotic" algorithms output for Cameraman image (the same image on all faces).

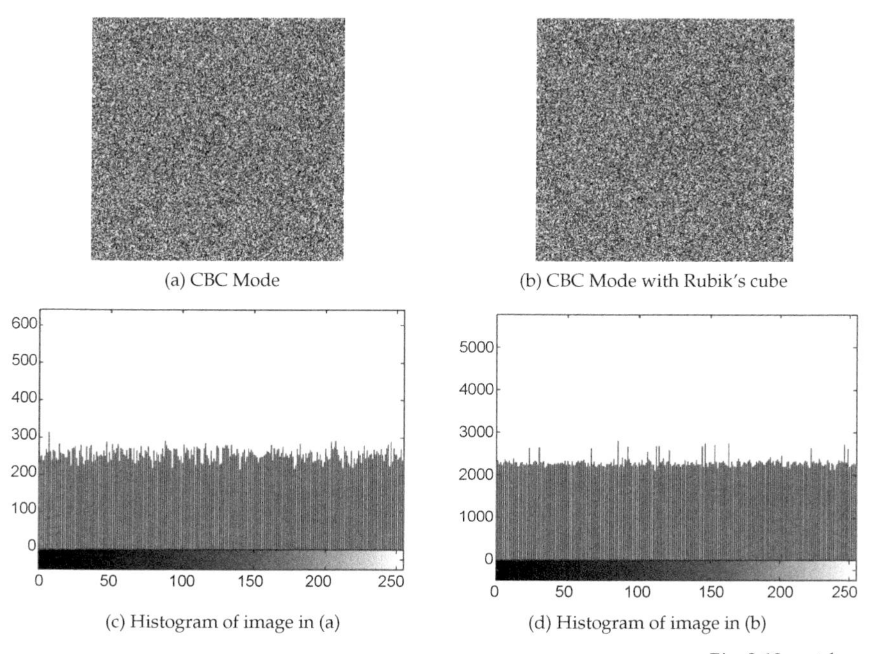

(a) CBC Mode

(b) CBC Mode with Rubik's cube

(c) Histogram of image in (a)

(d) Histogram of image in (b)

Fig. 3.13 contd. ...

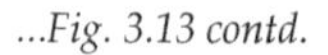

...Fig. 3.13 contd.

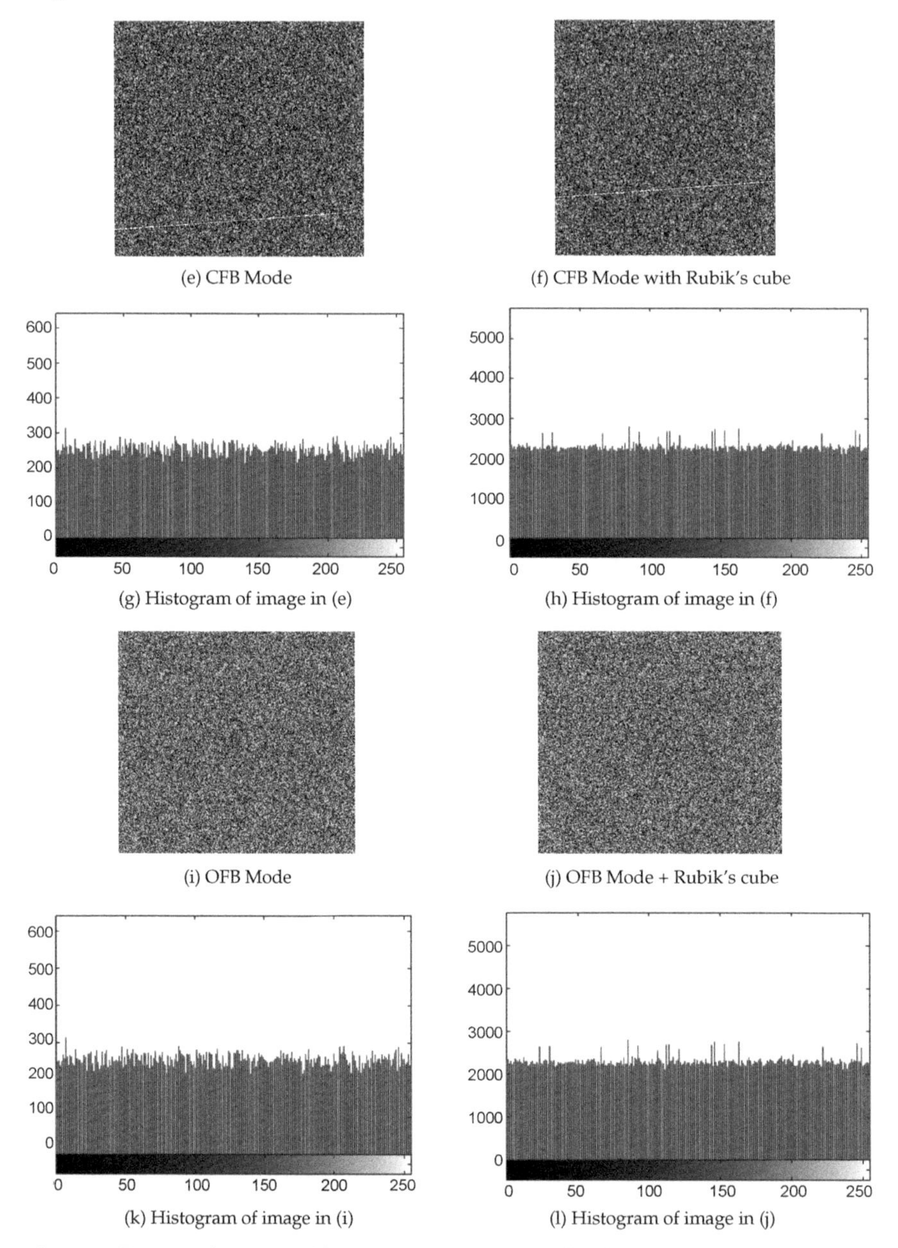

(e) CFB Mode

(f) CFB Mode with Rubik's cube

(g) Histogram of image in (e)

(h) Histogram of image in (f)

(i) OFB Mode

(j) OFB Mode + Rubik's cube

(k) Histogram of image in (i)

(l) Histogram of image in (j)

Fig. 3.13: Encrypted image and histogram of "Chaotic and Rubik's cube" algorithm output vs. "Chaotic" algorithm output for Mohamed image (similar images on all faces).

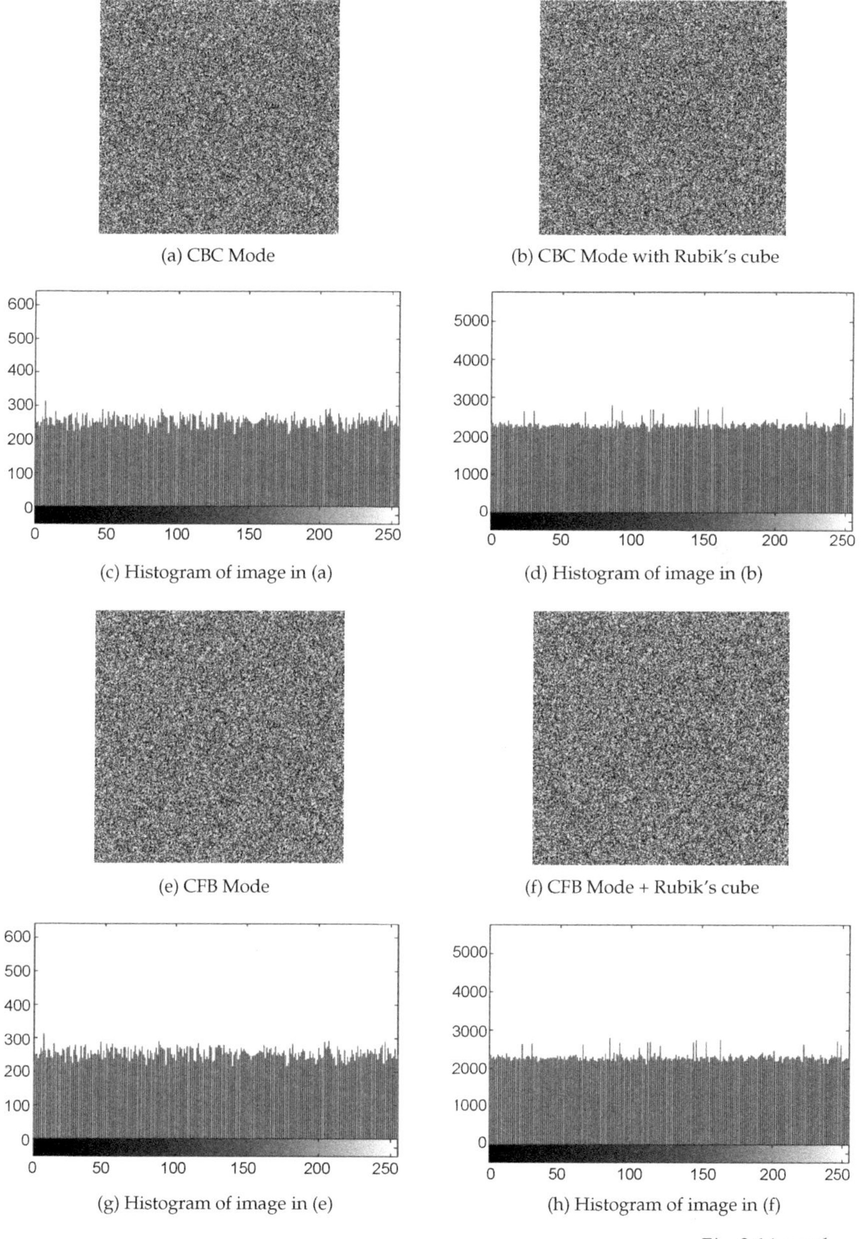

(a) CBC Mode

(b) CBC Mode with Rubik's cube

(c) Histogram of image in (a)

(d) Histogram of image in (b)

(e) CFB Mode

(f) CFB Mode + Rubik's cube

(g) Histogram of image in (e)

(h) Histogram of image in (f)

Fig. 3.14 contd. ...

...Fig. 3.14 contd.

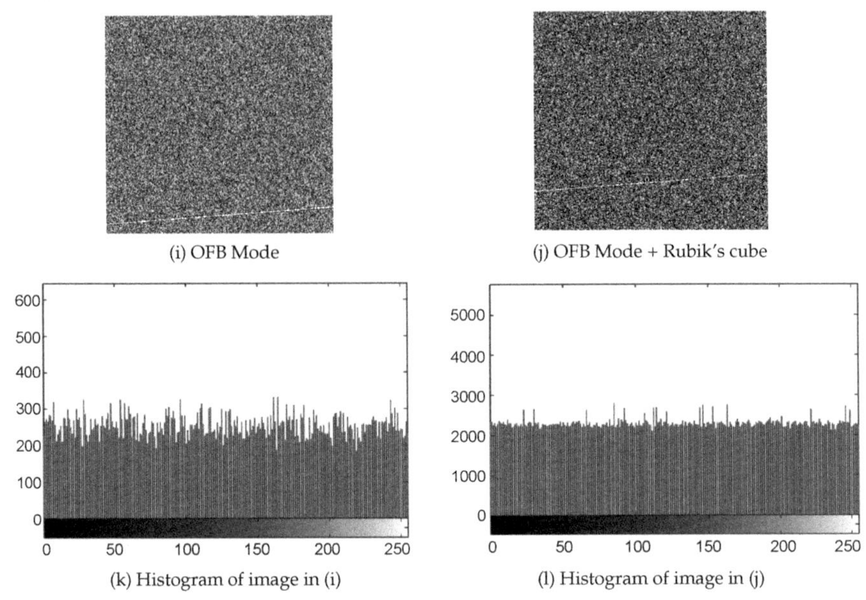

(i) OFB Mode

(j) OFB Mode + Rubik's cube

(k) Histogram of image in (i)

(l) Histogram of image in (j)

Fig. 3.14: Encrypted image and histogram of "Chaotic and Rubik's cube" algorithm output vs. "Chaotic" algorithm output for Cameraman and Mohamed images (different images on all faces).

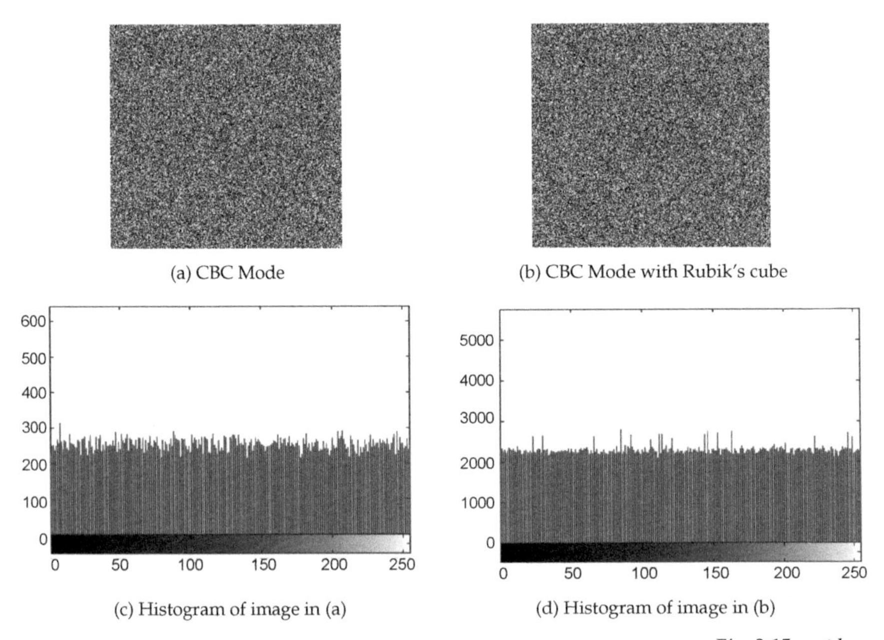

(a) CBC Mode

(b) CBC Mode with Rubik's cube

(c) Histogram of image in (a)

(d) Histogram of image in (b)

Fig. 3.15 contd. ...

...Fig. 3.15 contd.

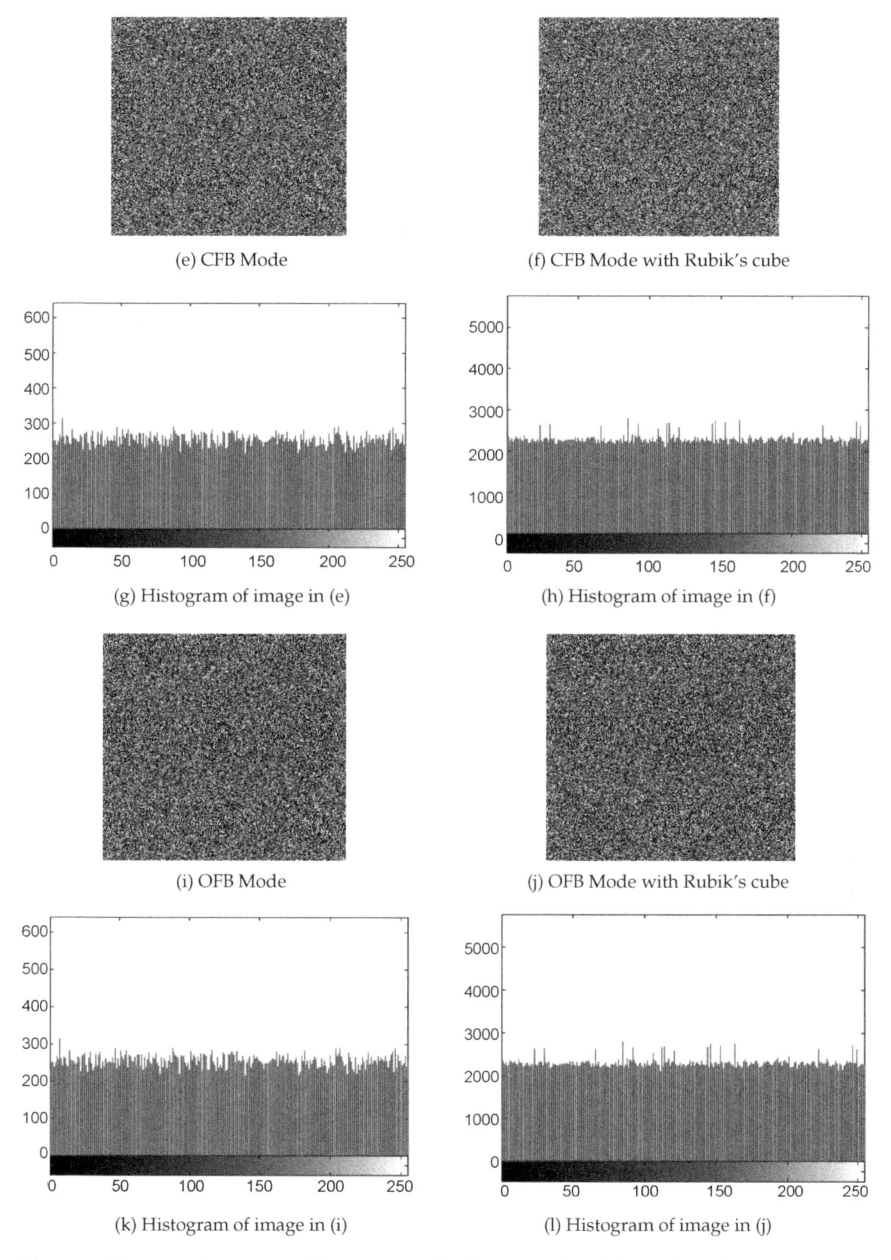

(e) CFB Mode

(f) CFB Mode with Rubik's cube

(g) Histogram of image in (e)

(h) Histogram of image in (f)

(i) OFB Mode

(j) OFB Mode with Rubik's cube

(k) Histogram of image in (i)

(l) Histogram of image in (j)

Fig. 3.15: Encrypted image and histogram of "Chaotic and Rubik's cube" algorithm output vs. "Chaotic" algorithm output for Cameraman and Mohamed images (different images on all faces).

3.5.2 Deviation

The largest two deviations are for CFB + Rubik's algorithm, then chaotic with OFB mode. As the deviation increases, the algorithm becomes strong.

Table 3.7: Deviation results for the different modes.

Encryption Algorithm	Deviation	
	Cameraman	Mohamed
CBC encryption	35.24	29.73
CBC + Rubik's cube encryption (Similar faces)	35.39	29.79
CBC + Rubik's cube encryption (Different faces)	35.52	29.87
CFB encryption	35.63	29.92
CFB + Rubik's cube encryption (Similar faces)	35.90	30.19
CFB + Rubik's cube encryption (Different faces)	35.72	30.21
OFB encryption	35.63	29.85
OFB + Rubik's cube encryption (Similar faces)	35.74	30.09
OFB + Rubik's cube encryption (Different faces)	35.74	30.09

3.5.3 Correlation Coefficient

The smallest three correlation values are for CBC + Rubik's algorithm, then CFB + Rubik's algorithm, and finally OFB + Rubik's algorithm. As the correlation decreases, the algorithm becomes stronger.

Table 3.8: Correlation results for the different algorithms.

Encryption Algorithm	Correlation	
	Cameraman	Mohamed
CBC encryption	0.0050	–0.0045
CBC + Rubik's cube encryption (Similar faces)	–0.0098	–0.0055
CBC + Rubik's cube encryption (Different faces)	–0.0038	–0.0058
CFB encryption	–0.0015	–0.0050
CFB + Rubik's cube encryption (Similar faces)	–0.0080	8.8958e-04
CFB + Rubik's cube encryption (Different faces)	–0.0067	–0.0068
OFB encryption	–0.0050	–0.0032
OFB + Rubik's cube encryption (Similar faces)	–0.0033	–0.0038
OFB + Rubik's cube encryption (Different faces)	–0.0033	–0.0038

3.5.4 Processing Time

The calculations recording that Rubik's algorithm needs more time in processing, as shown in Table 3.9.

Table 3.9: The processing time for the different modes/Cameraman.

Encryption Algorithm	Processing Time
CBC encryption	0.80 seconds
CBC + Rubik's cube encryption (Similar faces)	8.74 seconds
CBC + Rubik's cube encryption (Different faces)	10.03 seconds
CFB encryption	1.06 seconds
CFB + Rubik's cube encryption (Similar faces)	9.49 seconds
CFB + Rubik's cube encryption (Different faces)	10.06 seconds
OFB encryption	3.64 seconds
OFB + Rubik's cube encryption (Similar faces)	8.76 seconds
OFB + Rubik's cube encryption (Different faces)	11.24 seconds

3.5.5 Noise Immunity

The sensitivity of chaotic encryption with the different modes of operation to the BER is shown in Fig. 3.16 and Fig. 3.17. The chaotic encryption with OFB mode is more robust to noise than the other modes, while at large values of BER, the chaotic encryption with OFB mode + Rubik's cube is the best.

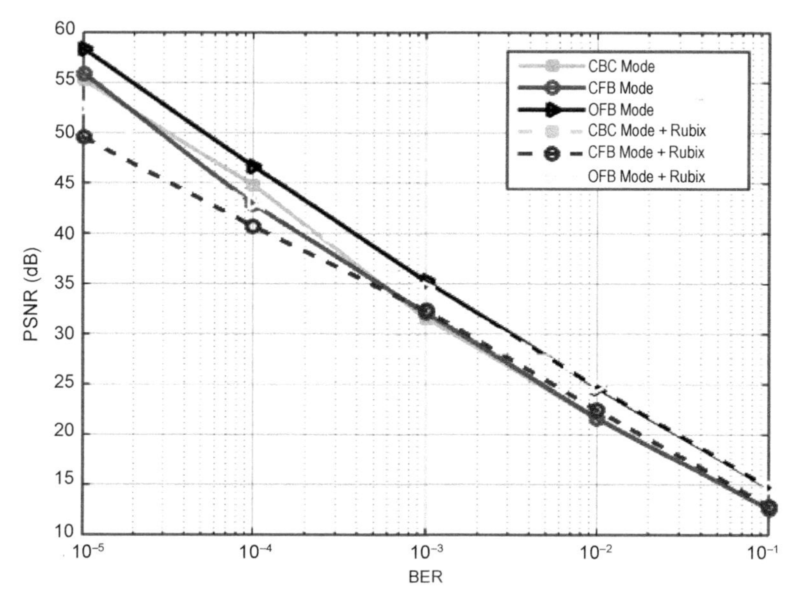

Fig. 3.16: PSNR vs. BER for encrypted image transmission with OFDM system.

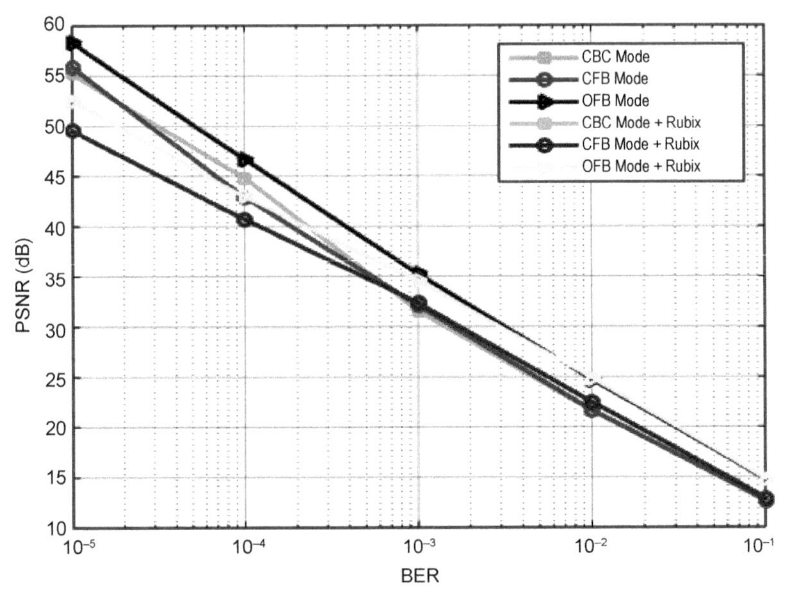

Fig. 3.17: PSNR vs. BER for encrypted image transmission with OFDM system.

3.6 Proposed Hybrid Encryption Framework

In this section, the suggested hybrid encryption framework for efficient transmission of images over OFDM system is introduced. The general structure of the proposed image communication system is shown in Fig. 3.18. The compressed image bit streams are encrypted using the proposed hybrid encryption framework, and then transmitted through the OFDM system. Then, the images are received, and decrypted.

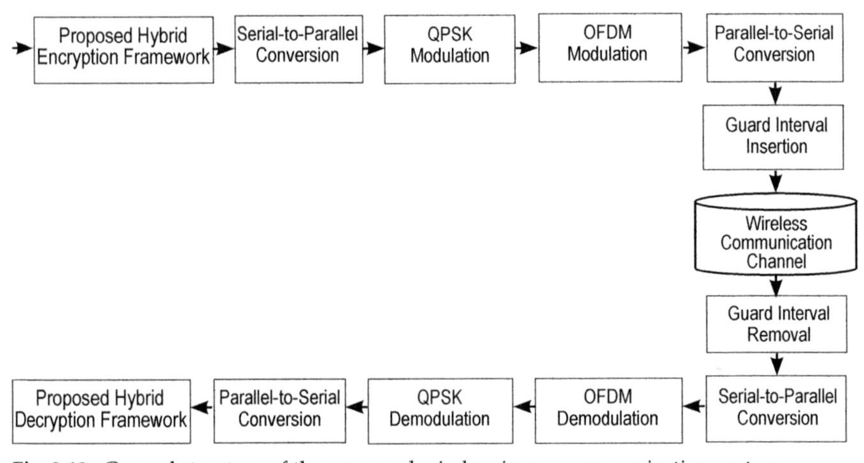

Fig. 3.18: General structure of the proposed wireless image communication system.

Figure 3.18 briefly describes the key steps in an OFDM communication system, which is used for the transmission of the encrypted images. The first step is the image encryption. Then, the encrypted images are transformed into a serial stream of bits. Thus, there is a need for a serial-to-parallel step to put the encrypted images data in a form valid for the digital modulation step [28]. In this way, the aggregate symbol rate is maintained, but each sub-carrier experiences flat fading, or Inter Symbol Interference (ISI)-free communication. Moreover, a guard interval exceeding the delay spread is used. The resulting symbols are sent in a serial manner through the wideband channel. At the receiver, the guard interval is discarded, and the received symbols are demodulated using FFT, DCT, or DWT [29].

3.7 Simulation Results

Simulation experiments have been performed to simulate the AES/RC6/Chaotic with Rubik's cube method. The encrypted images are digitized and transmitted via the OFDM modulation. Finally, the data is transferred through AWGN and Rayleigh fading channels. In all the simulation experiments, OFDM system is modulated by QPSK and 128 sub-channels, each with 500 kHz spacing. The summarized parameters are given in Table 3.10.

Table 3.10: System simulation parameters.

System Bandwidth	CP Length	Number of Sub-carriers	Sub-carrier Spacing	Type of Transform
64 MHz	16 samples	128	500 kHz	FFT, DCT, and DWT

The quality of the decrypted images is evaluated at the receiver through MATLAB®, via calculating the Peak Signal-to-Noise Ratio (PSNR), which is defined as follows:

$$PSNR = 10\log_{10}\left(\frac{255^2}{MSE}\right) \tag{5}$$

The Mean Square Error (MSE) is defined as:

$$MSE = \frac{1}{M^2}\sum_{i=1}^{M}\sum_{j=1}^{M}\left[h(i,j) - \hat{h}(i,j)\right]^2 \tag{6}$$

where $h(i, j)$ is the original image with dimensions $M \times M$ and $\hat{h}(i, j)$ is the decrypted image.

3.7.1 AES with Rubik's Cube Algorithm (Similar Faces)

Figure 3.19 illustrates the PSNR performance vs. the E_b/N_0 for the OFDM system with its three different versions via the AWGN channel for AES + Rubik's encryption applied on Cameraman image. The PSNR performances of the AES + Rubik cube encryption on Cameraman image via FFT/DWT are better than that without at $E_b/N_0 = 10$ db. The figure also shows that the PSNR performances of all algorithms have approximately the same behavior at range of $Eb/N_0 = [0–6]$ dB. At $E_b/N_0 = 8$ dB, the best is the DWT system.

Figure 3.20 illustrates the PSNR performance vs. the E_b/N_0 for the OFDM system over the Rayleigh fading channel at $f_d = 600$ Hz for AES + Rubik's encryption of Cameraman image. The PSNR performances of AES + Rubik's cube encryption over FFT-OFDM is the best at $E_b/N_0 = 10$ dB, while the AES without Rubik's cube encryption over FFT-OFDM scores a PSNR = 49.38 dB at $E_b/N_0 = 10$ dB as a maximum value recorded. The figure also shows that the PSNR performances of the different algorithms over DCT/DWT have approximately the same behavior at $E_b/N_0 = [0–6]$ dB.

Figure 3.21 illustrates the PSNR performance vs. E_b/N_0 for the FFT, the DCT, and the DWT systems with AES and AES + Rubik's cube encryption. The PSNR performances of AES + Rubik's cube encryption over FFT-OFDM are best at $E_b/N_0 = 10$ dB, where the received image is the same as that sent, while the AES algorithm without Rubik's cube algorithm over the

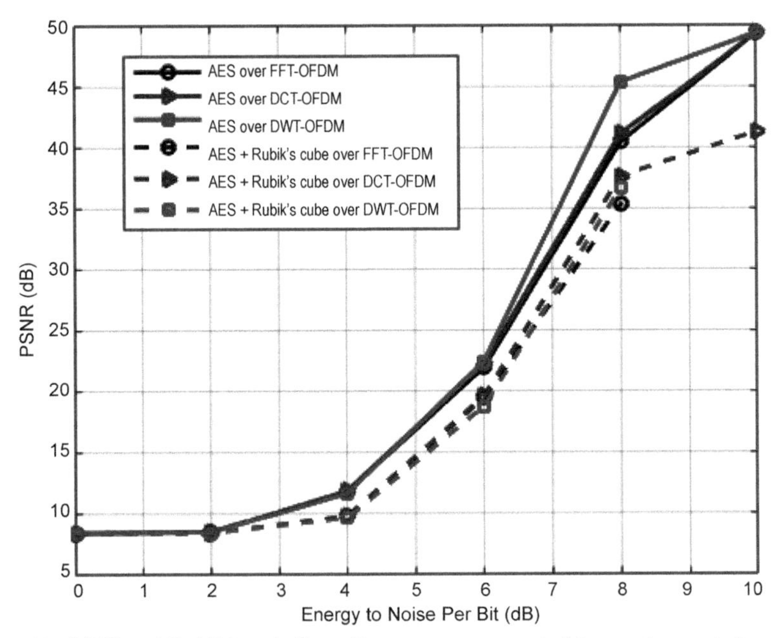

Fig. 3.19: "AES and Rubik's cube" on Cameraman encrypted image transmission over FFT/DCT/DWT system over AWGN channel.

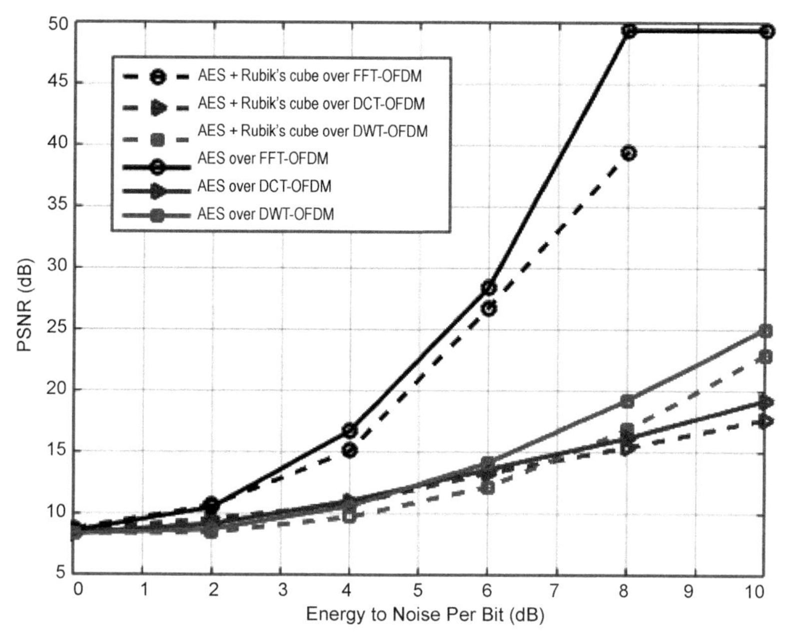

Fig. 3.20: "AES and Rubik's cube" encryption and transmission of Cameraman image via FFT/DCT/DWT system over the Rayleigh fading channel at $f_d = 600$ Hz.

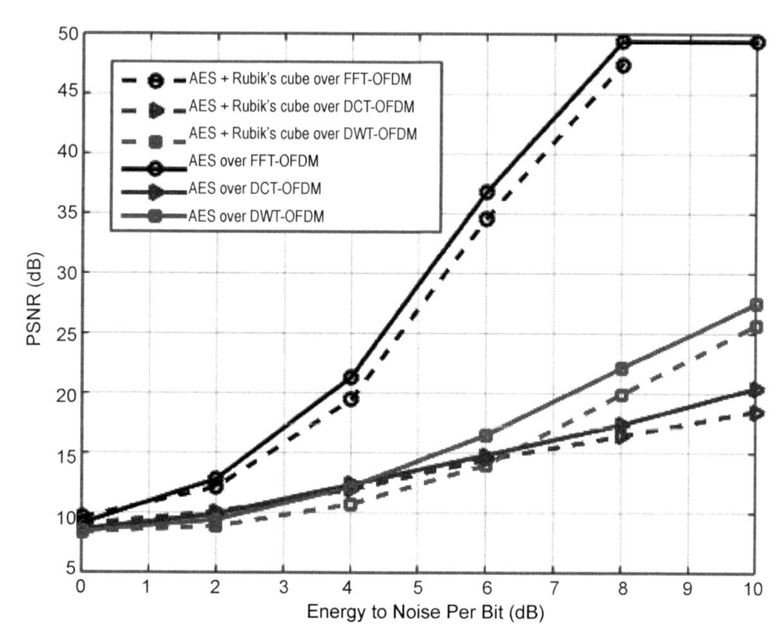

Fig. 3.21: PSNR vs. E_b/N_o for "AES and Rubik's cube" encrypted image transmission over FFT, DCT and DWT system for Rayleigh Fading, $f_d = 100$, with zero padding.

FFT-OFDM scores a PSNR = 49.38 dB at E_b/N_o = 10 dB. The figure shows that the PSNR performances of different algorithms via DCT/DWT have approximately the same behavior at E_b/N_o = [0–6] dB.

Simulation results show that the performance of the FFT-OFDM system through Rayleigh fading channel with AES algorithm in the presence of Rubik's cube is better than the AES algorithm only, especially at high E_b/N_o. Also, with FFT/DWT over AWGN channel, the performance gain is maximized to achieve perfect reconstruction at E_b/N_o = 10 dB. So, AES + Rubik's cube algorithm may be the more qualified for secure transmission, where Rubik's cube encryption provides more deviation, less correlation, suitable uniform histogram, and high qualified transmission of FFT-OFDM at E_b/N_o = 10 dB or more.

3.7.2 AES with Rubik's Cube Algorithm (Different Faces)

Figure 3.22 illustrates the PSNR performance vs. the E_b/N_o for FFT, DCT, and DWT systems via AWGN channel with AES + Rubik's cube encryption, the PSNR performances of AES + Rubik's cube encryption via the FFT/DCT/DWT channel are better than it without at E_b/N_o = 10 dB. The figure also shows that the PSNR performances of all algorithms have approximately the same behavior at the range of E_b/N_o = [0–6] dB. At E_b/N_o = 8 dB, the best is DWT system.

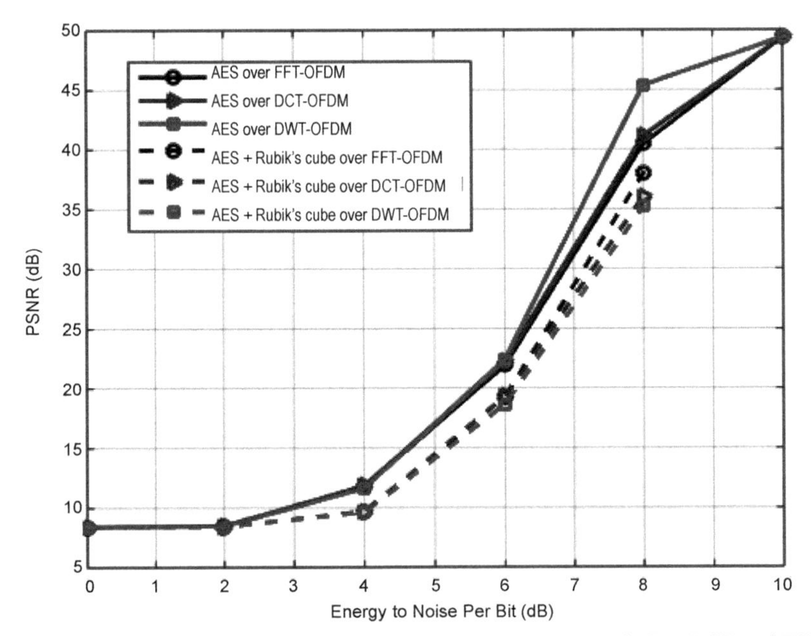

Fig. 3.22: PSNR vs. E_b/N_o for AES encrypted image transmission with FFT, DCT and DWT system over AWGN channel.

Figure 3.23 illustrates the PSNR performance vs. the E_b/N_o for the FFT, the DCT, and the DWT systems with AES and AES + Rubik's cube encryption. The PSNR performances of AES + Rubik's cube encryption over the FFT-OFDM are best at $E_b/N_o = 10$ dB, while the AES without the Rubik's cube algorithm via the FFT-OFDM scores a PSNR = 49.38 dB at $E_b/N_o = 10$ dB as a maximum value recorded. The figure also shows that the PSNR performances of the different algorithms via DCT/DWT-OFDM have approximately the same behavior at $E_b/N_o = [0–6]$ dB.

Figure 3.24 illustrates the PSNR performance vs. the E_b/N_o for the FFT, the DCT, and the DWT systems with AES and AES + Rubik cube encryption. The PSNR performances of AES + Rubik's cube over FFT-OFDM are best at $E_b/N_o = 8$ dB and 10 dB, where we receive the image as sent, while the AES without Rubik's cube algorithm via FFT-OFDM scores a PSNR = 49.38 dB. At $E_b/N_o = 10$ dB, we record a maximum value for this algorithm. The figure also shows that the PSNR performances of different algorithms via DCT/DWT have approximately the same behavior at $E_b/N_o = [0\ 06]$ dB.

Simulation results show that the performance of the FFT-OFDM system through Rayleigh fading channel with AES mode in the presence of Rubik's cube is better than with the AES algorithm only, especially at high E_b/N_o. Also, with FFT/DCT/DWT over AWGN channel, the performance gain is

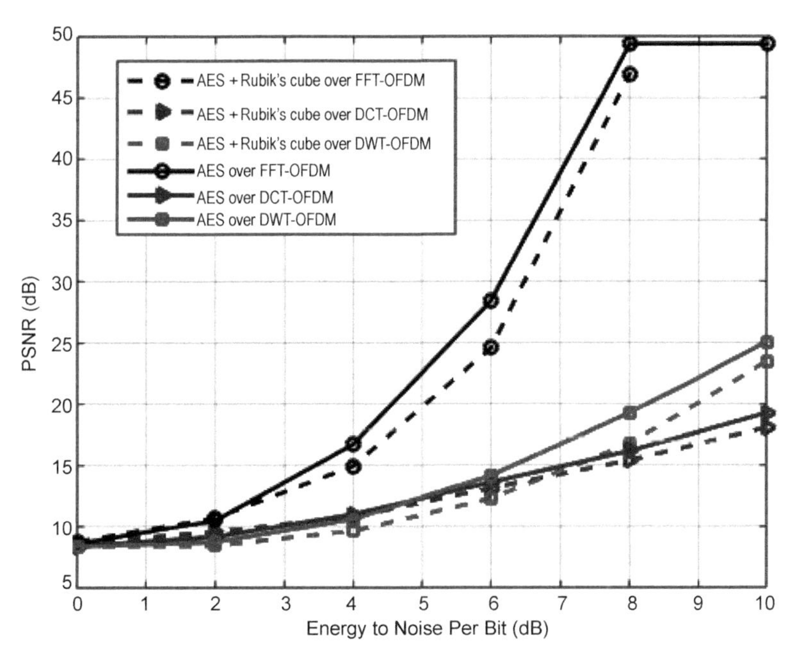

Fig. 3.23: PSNR vs. E_b/N_o for "AES and Rubik's cube" encrypted image transmission with FFT, DCT and DWT system for Rayleigh fading at $f_d = 600$ Hz.

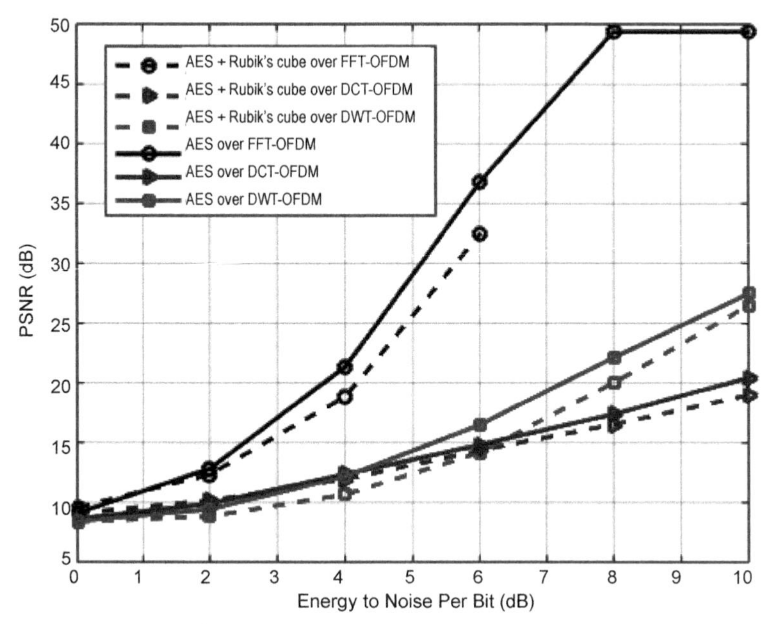

Fig. 3.24: PSNR vs. E_b/N_o for "AES and Rubik's cube" encrypted image transmission with FFT, DCT and DWT system for Rayleigh fading at f_d = 100 Hz.

maximized to record a PSNR = Inf at E_b/N_o = 10 dB. So, AES + Rubik's cube may be the most qualified for secure transmission, where Rubik's cube provides more deviation, less correlation, suitable uniform histogram and high qualified transmission of FFT-OFDM at E_b/N_o = 10 dB or more.

3.7.3 RC6 with Rubik's Cube Algorithm (Similar Faces)

Figure 3.25 illustrates the PSNR performance vs. the E_b/N_o for the OFDM system with its three different versions via the AWGN channel for Cameraman image implementing RC6 + Rubik's encryption. The PSNR performance of RC6 + Rubik's cube encryption via FFT/DCT channel is better than that without Rubik's cube at E_b/N_o = 10 dB. The figure also shows that the PSNR performances of all algorithms have approximately the same behavior in the range of E_b/N_o = [0–6] dB.

Figure 3.26 illustrates the PSNR performance vs. the E_b/N_o for the OFDM system with its three versions over AWGN channel with RC6 + Rubik's encryption. The PSNR performances for Lena image encryption with RC6 + Rubik's cube and transmission via FFT/DWT channel are better than that without Rubik's cube at E_b/N_o = 10 dB, where the decrypted image is received typically similar to the original image. Also, the figure shows that the PSNR of all algorithms have approximately the same behavior in the range of E_b/N_o = [0–6] dB.

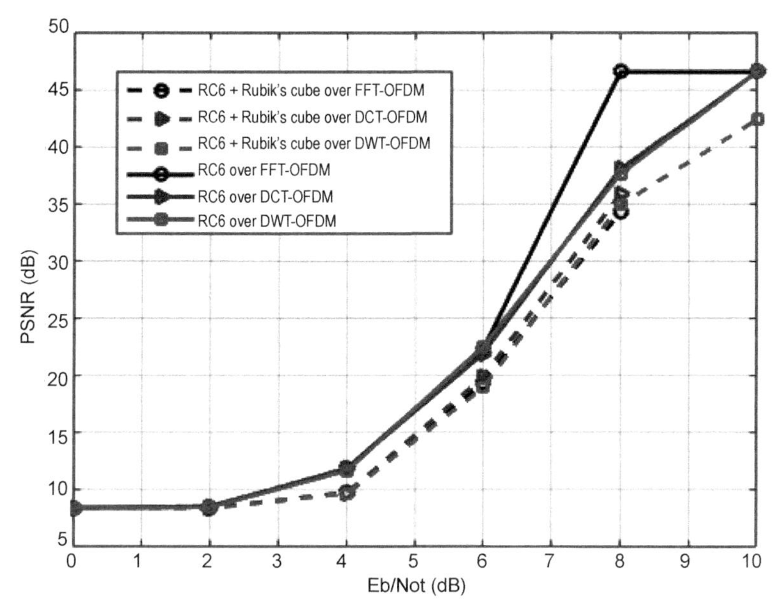

Fig. 3.25: PSNR vs. E_b/N_o for "RC6 and Rubik's cube" encryption for Cameraman image transmission with FFT, DCT and DWT system over AWGN channel.

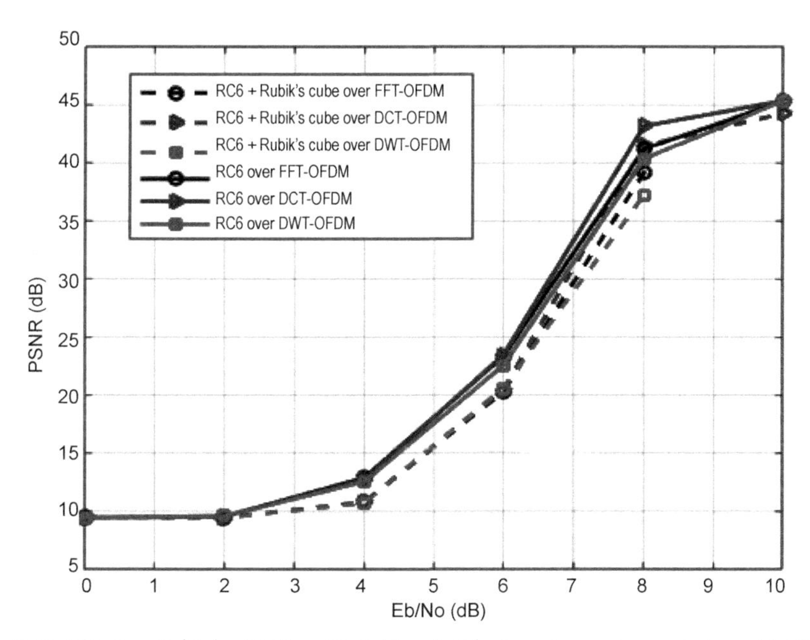

Fig. 3.26: PSNR vs. E_b/N_o for "RC6 and Rubik's cube" for Lena image transmission with FFT-OFDM, DCT-OFDM and DWT-OFDM system over AWGN channel.

Figures 3.27 to 3.29 illustrate the PSNR performance vs. the E_b/N_o for the OFDM system over the Rayleigh fading channel at f_d = 600 Hz for

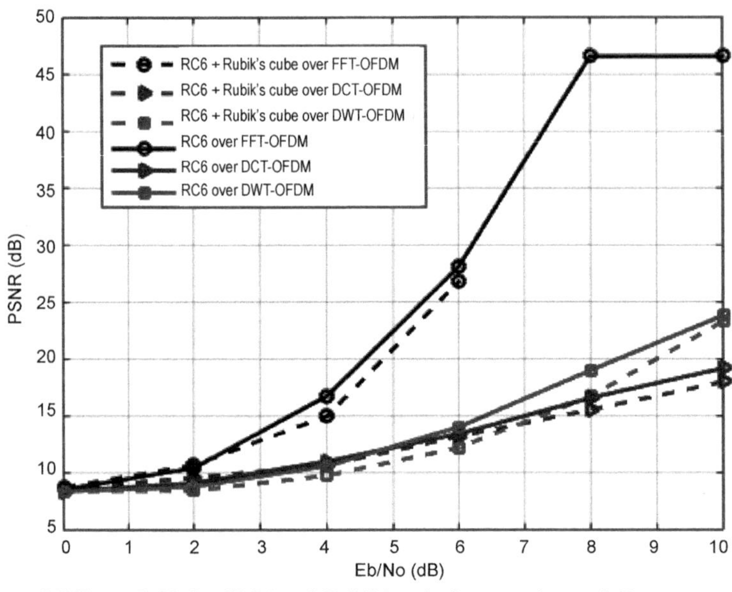

Fig. 3.27: PSNR vs. E_b/N_o for "RC6 and Rubik's cube" encryption and Cameraman image transmission with FFT, DCT and DWT systems for Rayleigh fading at f_d = 600 Hz.

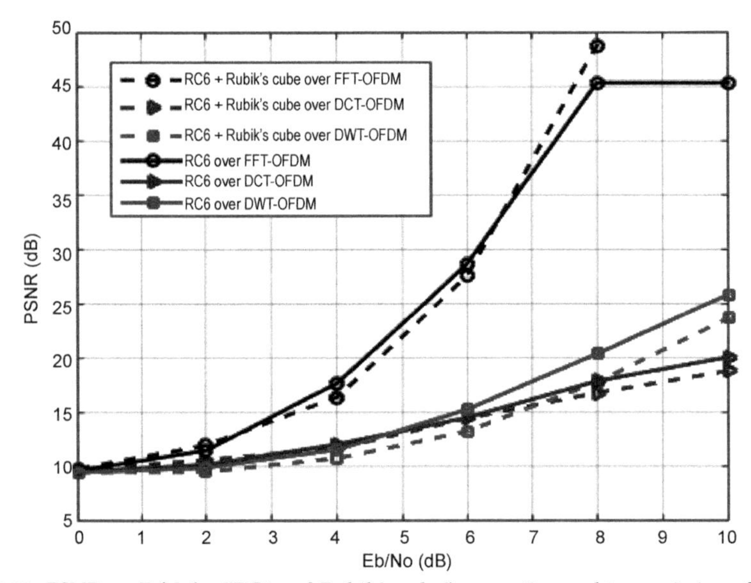

Fig. 3.28: PSNR vs. E_b/N_o for "RC6 and Rubik's cube" encryption and transmission of Lena image over FFT, DCT and DWT system for Rayleigh fading at f_d = 600 Hz.

Fig. 3.29: RC6 + Rubik's encrypted image transmission over Rayleigh Fading at f_d = 600 Hz at E_b/N_o = 10.

Cameraman and Lena images with RC6 + Rubik's encryption. The PSNR performances of RC6 + Rubik's cube encryption over FFT-OFDM are best at E_b/N_o = 10 dB, while RC6 without Rubik's cube encryption over FFT-OFDM scores a of PSNR = 46.6 dB for Cameraman, PSNR = 41.23 dB for

Mohamed, and PSNR = 45.3784 dB for Lena at E_b/N_o = 10 dB. This is the maximum value recorded before for this algorithm. The figure also shows that the PSNR performances of different algorithms via DCT/DWT have approximately the same behavior at E_b/N_o = [0–6] dB.

Simulation results show that the performance of the FFT-OFDM system over AWGN and Rayleigh fading channels with RC6 in the presence of Rubik's cube encryption is better than with only the RC6 algorithm, especially at high E_b/N_o. This hybrid algorithm is more qualified for secure transmission. It always provides more deviation, less correlation, suitable uniform histogram, and highly qualified transmission of FFT-OFDM at E_b/N_o = 10 dB or more.

3.7.4 RC6 with Rubik's Cube Algorithm (Different Faces)

Figure 3.30 illustrates the PSNR performance vs. E_b/N_o for the OFDM system with its three different versions over AWGN channel for Cameraman image encryption with RC6 + Rubik's cube. The PSNR performances of Cameraman for RC6 with Rubik's cube method via FFT/DCT/DWT channel are better than those without Rubik's cube at E_b/N_o = 10 dB. The figure also shows that the PSNR performances of all algorithms have approximately the same behavior at the range of E_b/N_o = [0–6] dB.

Figure 3.31 illustrates the PSNR performance vs. the E_b/N_o for the OFDM system with its three versions over AWGN channel for RC6 + Rubik's

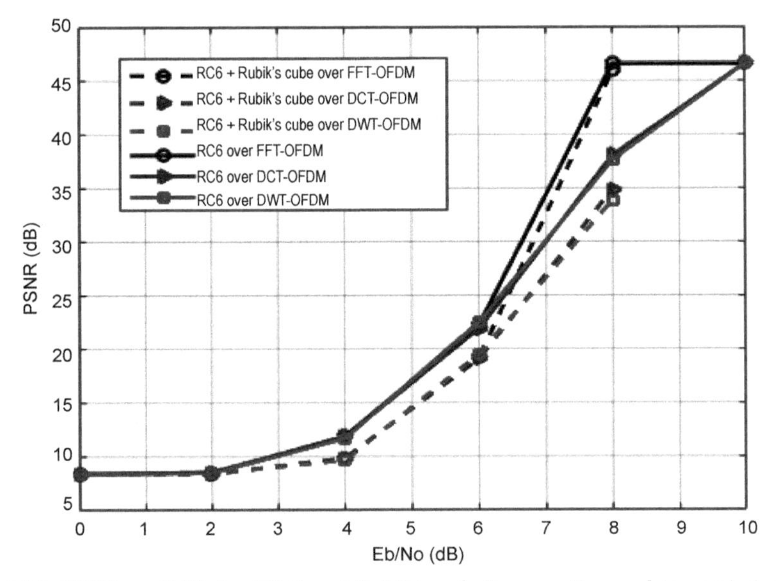

Fig. 3.30: PSNR vs. E_b/N_o for "RC6 and Rubik's cube" encryption and transmission of Cameraman image via FFT, DCT, and DWT systems over AWGN channel.

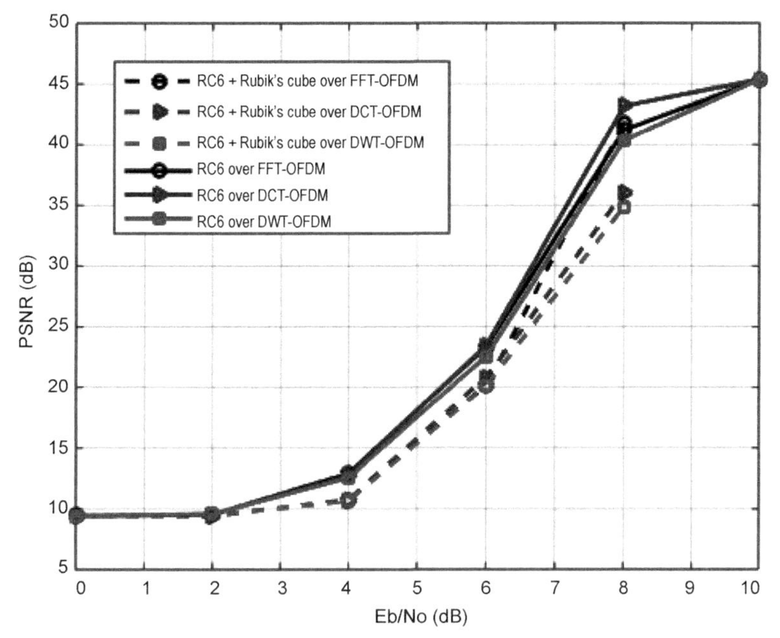

Fig. 3.31: PSNR vs. E_b/N_o for "RC6 and Rubik's cube" Lena image encryption and transmission via FFT, DCT, and DWT system over AWGN channel.

cube encryption. The PSNR performances for Lena image encrypted with RC6 + Rubik's cube via FFT/DCT/DWT channel are better than that without Rubik's cube at E_b/N_o = 10 dB, where the decrypted image is received typically similar to the original image. Also, the figure shows that the PSNR of all algorithms have approximately the same behavior in the range of E_b/N_o = [0–6] dB.

Figures 3.32 to 3.34 illustrate the PSNR performance vs. E_b/N_o for the OFDM system over the Rayleigh fading channel at f_d = 600 Hz for Cameraman and Lena image encryption and transmission with RC6+Rubik's cube. The PSNR performances of RC6 + Rubik's cube over FFT-OFDM are best at E_b/N_o = 10 dB and at 8 dB for Cameraman image, while RC6 without Rubik's cube via FFT-OFDM gives a PSNR = 46.6 dB for Cameraman, and PSNR = 45.38 dB for Lena at E_b/N_o = 10 dB as a maximum value recorded. The figure also shows that the PSNR performances of the different algorithms via DCT/DWT have approximately the same behavior at E_b/N_o = [0–6] dB.

Simulation results show that the performance of FFT/DCT/DWT system via AWGN channel with RC6 in the presence of Rubik's cube encryption is better than with only the RC6 algorithm, especially at high E_b/N_o. For Rayleigh fading channels, the performance of the FFT-OFDM system with RC6 encryption in the presence of Rubik's cube is better than with only the RC6 algorithm, especially at an E_b/N_o equal or

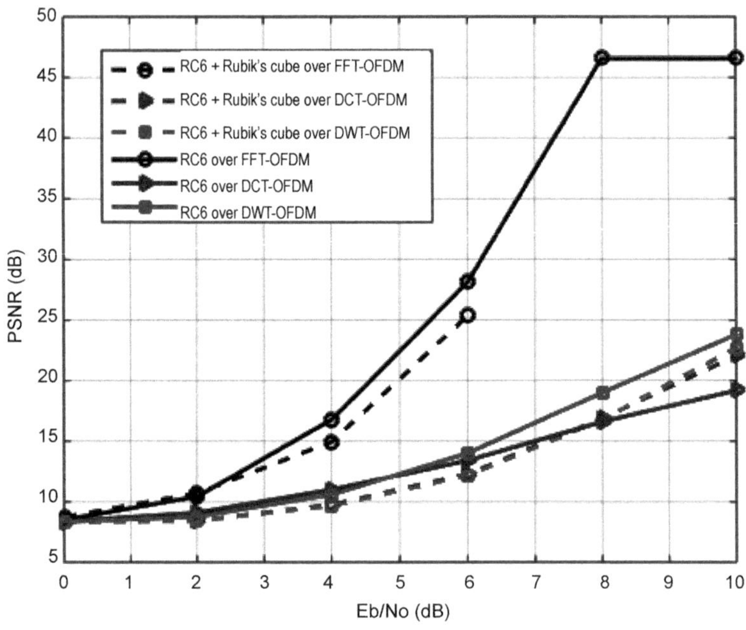

Fig. 3.32: PSNR vs. E_b/N_o for "RC6 and Rubik's cube" encryption and transmission of Cameraman image with FFT, DCT and DWT over Rayleigh fading channel at $f_d = 600$ Hz.

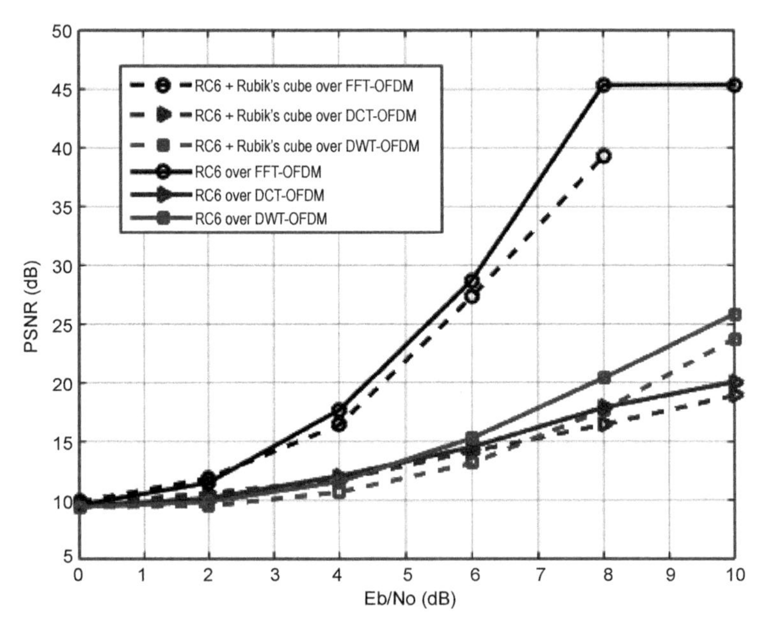

Fig. 3.33: PSNR vs. E_b/N_o for "RC6 and Rubik's cube" encryption and transmission of Lena image via FFT, DCT, and DWT system over Rayleigh fading channel at $f_d = 600$ Hz.

Fig. 3.34: RC6 and Rubik's cube encrypted image transmission over Rayleigh fading channel at f_d = 600 Hz at E_b/N_o = 10 dB.

higher than 10 dB. This hybrid algorithm is more qualified for secure transmission. It always provides more deviation, less correlation, suitable uniform histogram, and highly qualified transmission of FFT-OFDM at E_b/N_o = 10 dB or more.

3.7.5 Chaotic with Rubik's Cube Algorithm (Similar Faces)

Figure 3.35 illustrates the PSNR performance vs. the E_b/N_o for the FFT, the DCT, and the DWT systems with chaotic encryption implemented with CBC mode. The PSNR performances of all Rubik's cube-based algorithms are better than those without Rubik's cube at E_b/N_o = 10 dB. The figure also shows that the PSNR performances of all algorithms have the same behavior at E_b/N_o = [0–6] dB. At E_b/N_o = 8 dB, the best is the DCT system, then the DWT system.

Figure 3.36 illustrates the PSNR performance vs. the E_b/N_o for the FFT, the DCT, and the DWT systems with chaotic encryption implemented with CFB mode. The PSNR performances of Rubik's cube encryption and transmission via DCT-OFDM is the best at E_b/N_o = 8 dB, while all algorithms score an infinite PSNR at E_b/N_o = 10 dB. The figure also shows that the PSNR performances of all algorithms have approximately the same behavior at E_b/N_o = [0–6] dB.

Figure 3.37 illustrates the PSNR performance vs. the E_b/N_o for the FFT, the DCT, and the DWT systems with chaotic encryption implemented with OFB mode. The PSNR performance of DCT is the best at E_b/N_o = 8 dB, while all algorithms achieve an infinite PSNR = Inf at E_b/N_o = 10 dB. The figure also shows that the PSNR performances of Rubik's cube with FTT-OFDM and DWT are best at E_b/N_o = 8 dB.

Simulation results show that the performance of the OFDM system with chaotic modes in the presence of Rubik's cube encryption is better than

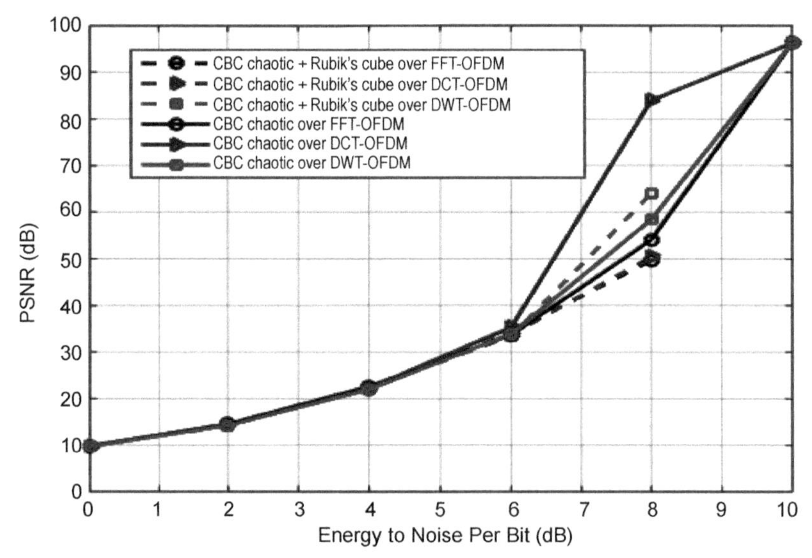

Fig. 3.35: PSNR vs. E_b/N_o for chaotic encryption CBC mode of Cameraman image and then transmission with FFT, DCT and DWT over Rayleigh fading channel.

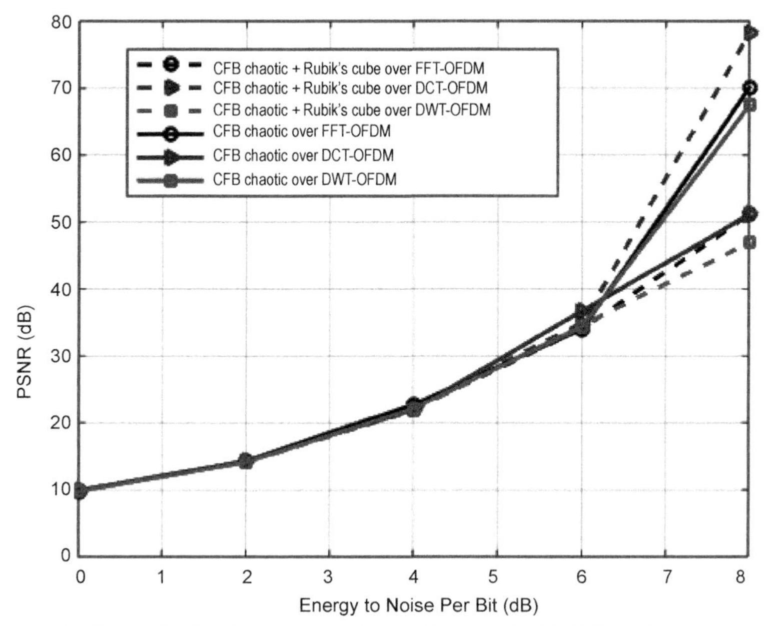

Fig. 3.36: PSNR vs. E_b/N_o for chaotic encryption implemented with CFB mode for Cameraman image transmission via FFT, DCT and DWT systems over Rayleigh fading channel.

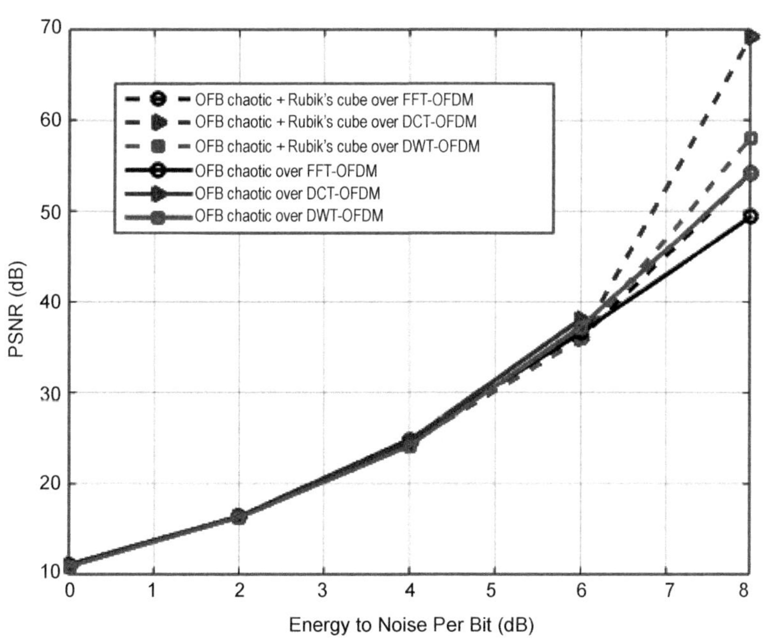

Fig. 3.37: PSNR vs. E_b/N_o for chaotic encryption implemented with OFB mode for Cameraman image transmission via FFT, DCT, and DWT systems over Rayleigh fading channel.

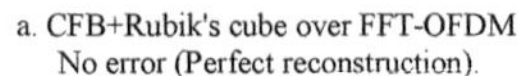

a. CFB+Rubik's cube over FFT-OFDM No error (Perfect reconstruction).

b. CFB+Rubik's cube over DCT-OFDM No error (Perfect reconstruction).

c. CFB+Rubik's cube over DWT-OFDM. PSNR = 47.3255dB

Fig. 3.38: "CFB and Rubik's cube" encryption for Cameraman image transmission via Rayleigh fading channel at f_d = 600 Hz at E_b/N_o = 10 dB.

with only the chaotic algorithm, especially at high E_b/N_o. Also, with DCT, the performance gain is about 30dB for the chaotic + CFB mode+ Rubik's as compared with the FFT and DWT at E_b/N_o = 8 dB. So, Rubik's cube encryption may be the more qualified for secure image transmission, where Rubik's cube provides more deviation, less correlation, suitable uniform histogram and more resistance to noise, which appears at high BERs.

3.7.6 Chaotic with Rubik's Cube Algorithm (Different Faces)

Figure 3.39 illustrates the PSNR performance vs. the E_b/N_o for the FFT, the DCT, and the DWT systems with chaotic of CBC mode, the PSNR performances of all Rubik's are better than that without at E_b/N_o = 10 dB. The Figure also shows that the PSNR performances of all algorithms have the same behavior at E_b/N_o = [0–6] dB. At E_b/N_o = 8 dB the best is DCT system.

Figure 3.40 illustrates the PSNR performance vs. E_b/N_o for the FFT-OFDM, the DCT, and the DWT systems with chaotic encryption implemented in the CFB mode. The PSNR performances of Rubik's cube encryption and transmission via DCT are better at E_b/N_o = 8 dB than those of chaotic encryption and transmission via DCT, while all algorithms score a PSNR = Inf at E_b/N_o = 10 dB. The figure also shows that the PSNR performances of all algorithms have approximately the same behavior at E_b/N_o = [0–6] dB.

Figure 3.41 illustrates the PSNR performance vs. the E_b/N_o for the FFT, the DCT, and the DWT systems with chaotic encryption implemented with OFB mode. The PSNR performances of DCT are best at E_b/N_o = 8 dB, while all algorithms provide a PSNR = Inf at E_b/N_o = 10 dB. The figure also shows that the PSNR performances of Rubik's cube encryption with FTT and DWT algorithms are best at E_b/N_o = 8 dB.

Simulation results show that the performance of the OFDM system with chaotic encryption modes in the presence of Rubik's cube is better than with only the chaotic algorithm, especially at high E_b/N_o. Also, with DCT,

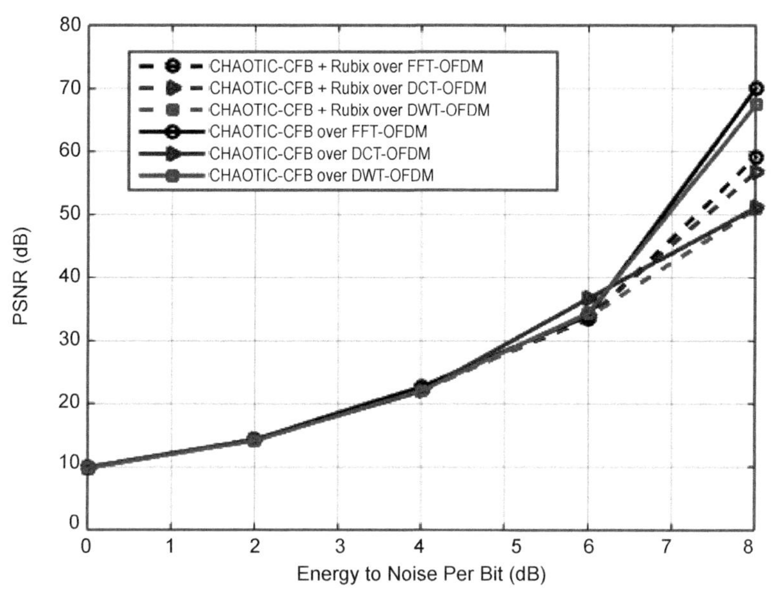

Fig. 3.39: PSNR vs. E_b/N_o for chaotic encryption implemented with CFB mode for Cameraman image transmission via FFT, DCT and DWT systems.

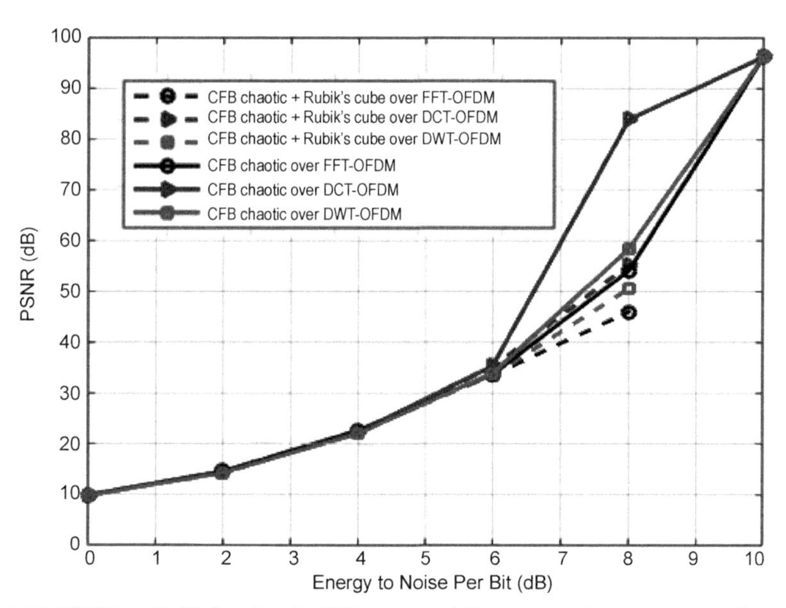

Fig. 3.40: PSNR vs. E_b/N_o for chaotic-CFB encrypted Cameraman image transmission with FFT, DCT and DWT system for Rayleigh fading.

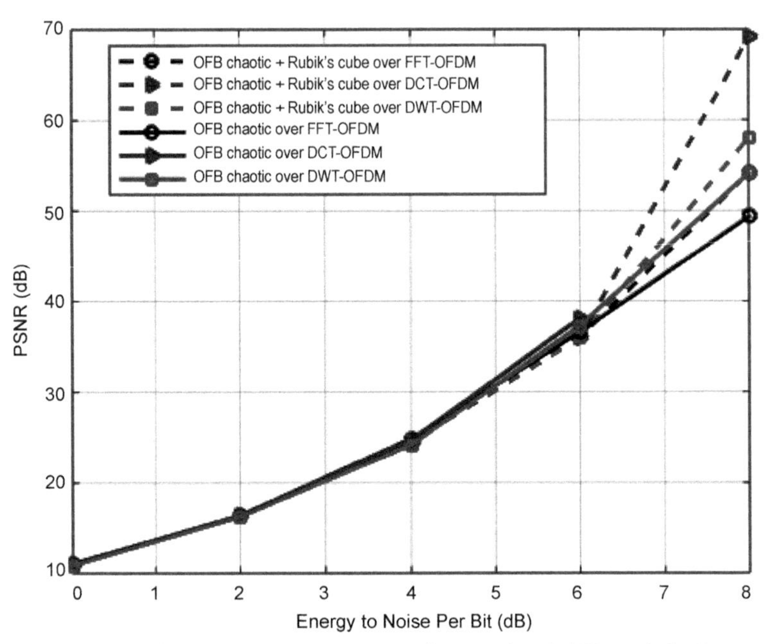

Fig. 3.41: PSNR vs. E_b/N_o for chaotic encryption implemented with OFB mode for Cameraman image transmission via FFT, DCT, and DWT system over Rayleigh fading channel.

the performance gain is about 30 dB for the CFB + chaotic + Rubik's cube algorithm as compared with the FFT-OFDM and DWT at $E_b/N_o = 8$ dB. So, Rubik's cube may be more qualified for secure transmission, where Rubik's cube provides more deviation, less correlation, suitable uniform histogram and more resistance to noise, which appears at high BERs.

Chapter 4

Hybrid Rubik's Cube Algorithm for Wireless Communications

4.1 Introduction

In this chapter, a hybrid encryption framework based on the Rubik's cube is suggested as a means to attain simultaneous encryption of a group of images. The suggested hybrid encryption framework begins with CFB operation mode of chaotic Baker map permutation or AES or RC6 technique as a first step for encrypting the multiple images, separately. Then, the resulting encrypted images are further encrypted in a second stage with Rubik's cube technique. Chaotic, RC6, or AES encrypted images are utilized as the faces of the Rubik's cube. From the concepts of image encryption, the RC6 or AES technique introduces a degree of diffusion, whilst the chaotic Baker map adds a degree of permutation. Moreover, the Rubik's cube technique adds more permutation to the encrypted images, simultaneously. The encrypted images are further transported through a wireless channel with different versions of OFDM system, and decrypted at the receiver side. The quality evaluation of the decrypted images at the receiver side reveals good performance. The simulation results reveal that the suggested hybrid encryption framework is efficient, and it presents strong security and robustness.

The standard of Multi-view Image Compression (MIC) [74, 75] attains efficient image encoding. It has obtained a lot of attention recently. The MIC standard is a complement of the 2-D image compression standard [76], and it is predicted to quickly replace the 2-D images in various implementations like education, medicine, entertainment, and gaming. The 3-D image communication through wireless networks has increased, dramatically [77, 78]. To transmit images over limited-resources channels, a strong enough encoding process must be utilized, whilst maintaining an appreciated image quality at the receiver. For efficient image communication, the 3-D-MIC system must exploit the spatial and temporal matching among neighboring areas in the same image in addition to the inter-view correlation inside the various image sequences in order to enhance the encoding efficiency. However, image compression with high rates is more sensitive to transmission channel corruptions. Image transmission through wireless channels is permanently subject to packet errors of both burst and random natures [79, 82].

The need for high speed and secure wireless multimedia communication systems has grown in the last few decades. The progress in mobile communications, satellite applications, Internet applications, and computer networks has given rise to new problems regarding security and privacy [83]. Having a reliable and secure means for transmission of multimedia is a necessary issue. Hence, data encryption and network security have become significant [84]. Multimedia encryptions have more implementations in different fields, such as medical imaging, internet communication, telemedicine, multimedia systems, and military communications [85]. Therefore, because of the fast progress in network development, humans can easily and arbitrarily distribute or access digital multimedia data from networks. The ownership security has become a significant issue for individuals, and it requires more attention. Thus, there is a significant threat to copyright owners and digital multimedia producers to conserve multimedia from intruder prospection to avoid losses in transmitted data [86–88]. Encryption is one of the most favorable methods to secure digital multimedia files in the domains of copyright protection and data authentication.

The objective of this chapter is to introduce an efficient hybrid image encryption framework from the security perspective for the images to be transported through a wireless channel, while preserving the good quality of the received images. We have adopted the OFDM as the transmission system for wireless communication of the encrypted images. Different versions of OFDM, such as FFT, DCT, and DWT, are studied in this work in order to choose the most convenient modulation technique that can achieve our objectives of high security and good quality of received images. Various channel impairments, such as AWGN and fading, are considered in this chapter. Moreover, channel equalization is investigated in order to study its effect on the quality of the received image images. So, we concentrate in our study on the ability of the suggested hybrid encryption framework to deal with channel degradations. In the proposed hybrid encryption framework, we need to achieve both diffusion and permutation in the encrypted images. To this aim, we can use chaotic, RC6, AES technique in a pre-processing step to achieve the permutation and diffusion. The Rubik's cube is used afterwards to achieve a better degree of permutation.

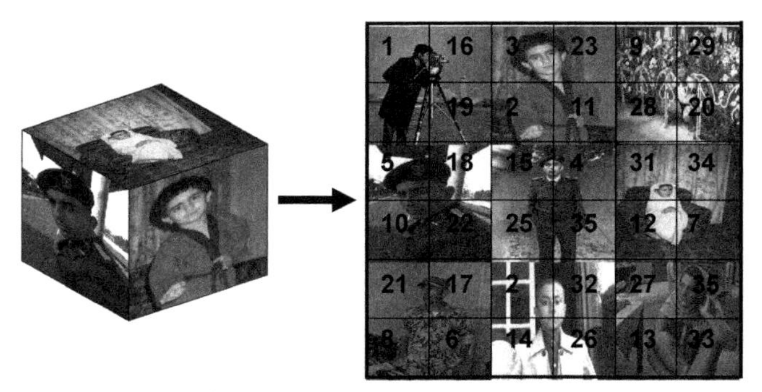

Fig. 4.1: Converting 3-D Rubik's cube into a 2-D image.

The proposed cryptosystem is based on Rubik's cube technique mixed with chaotic Baker map, RC6, and AES techniques. The IV of the chaotic maps works as the main key, and the block size is $N \times N$ pixels [89–91].

The Rubik's cube technique that will be adopted in this chapter in order to achieve further permutation in the encryption process has nine faces. It will be used to encrypt a group of 3-D images, simultaneously. The 3-D images implemented on these nine faces are those encrypted firstly with the chaotic, RC6, or AES technique [92–95]. The proposed hybrid encryption framework will guarantee both diffusion and permutation in the encrypted 3-D images. We will study its sensitivity to the wireless channel impairments, and also study the effect of channel equalization on the received image quality.

4.2 Proposed Hybrid Encryption Framework

The proposed hybrid encryption algorithm has three levels of encryption, as shown in Fig. 4.2 and Fig. 4.3:

1- **First degree** of encryption: The classic encryption algorithms, like chaotic, AES, and RC6.

2- **Second degree** of encryption: Rubik's cube algorithm applied on the output of the first stage.

3- **Third degree** of encryption: Arrangement of encryption algorithms used on the faces of the Rubik's cube as follows:

3-a) Chaotic/CFB - AES - RC6 - AES - RC6 - Chaotic/CFB - AES - RC6 - Chaotic/CFB.

3-b) Chaotic/CFC - RC6 - AES - RC6 - Chaotic/CFC - AES - RC6 - AES - Chaotic/CFB.

3-c) Chaotic/CFC - RC6 - AES - RC6 - Chaotic/CFC - AES - RC6 - AES - Chaotic/CFB.

The suggested hybrid encryption framework for efficient image transmission over OFDM system is introduced. Firstly, the huge amount

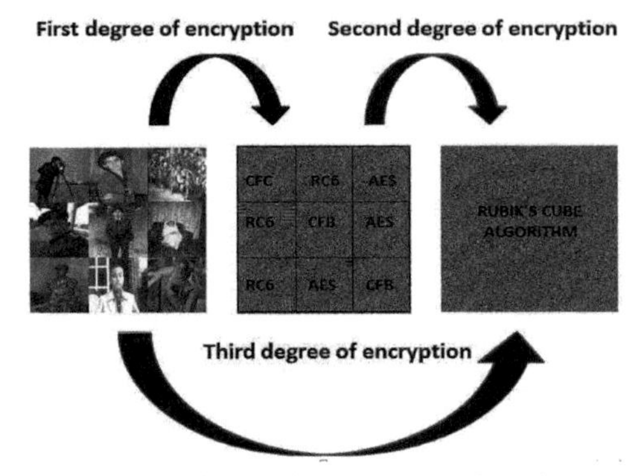

Fig. 4.2: The security degrees of Rubik's cube encryption algorithm.

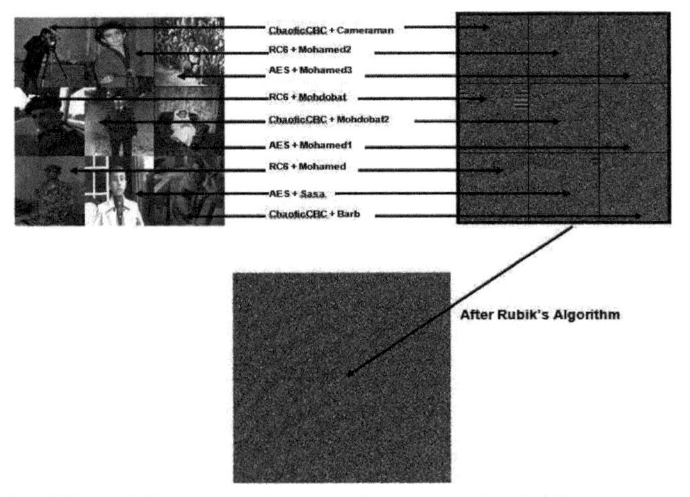

Fig. 4.3: The different 2-D images and encrypted versions using Rubik's cube encryption.

data of the image stream is compressed utilizing the encoder to reduce the image data size so as to be compatible with the transmission requirements over limited-resources wireless channels. After that, the compressed images bit streams are encrypted using the proposed hybrid encryption framework and transmitted through the OFDM system. Then, the image data are received and then decrypted.

In the proposed communication system, the images to be transmitted are firstly encrypted framework, as illustrated in Fig. 4.4. The encryption and communication processes can be summarized as follows:

Input the database of images of the database to encrypt;

- initialize all process;
- receive the image bit-streams;
- Start with chaotic, AES, and RC6 encryption algorithm in a defined key;

for all Encrypted images, ***do***
- Set the encrypted images over Rubik's cube nine faces;
- Convert the whole faces into a 2-D image with larger size;
- Randomize the new image with Rubik's cube mechanism;
- Transform the image into binary format;

end for

if Proposed Hybrid Encryption algorithm is ***done, then***
- Perform OFDM modulation;
- Transmit the signal over the wireless channel;
- Perform channel equalization at the receiver;
- Transform the received signal into binary format;
- Reconstruct the image into pixel values;

end if

for all Received encrypted images, ***do***
- Return to the canonical basis to get the decrypted images;

end for

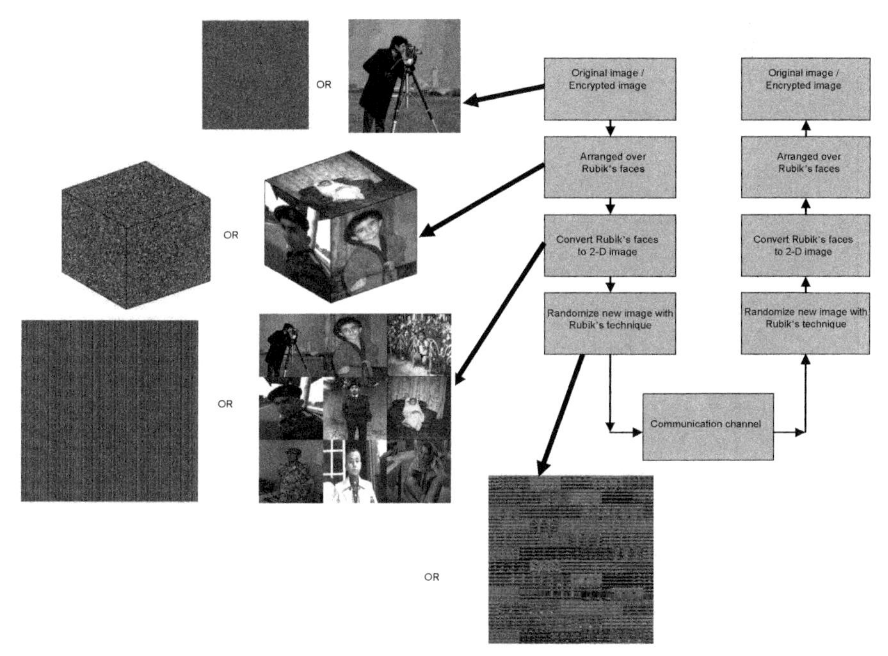

Fig. 4.4: The proposed hybrid encryption and communication framework.

4.3 Encryption Quality Evaluation Metrics

One of the important parameters in examining the encrypted 3-D images is the visual inspection, where the highly hidden features mean good encryption. Dependence on the visual inspection is not enough for encryption quality evaluation. So, different metrics are considered when evaluating the strength of the encryption algorithm [96, 99]. With the development of an encryption algorithm, different changes take place in pixel values as compared to the values before the image encryption. This means that the greater the change in pixel values, the more effective the image encryption will be, and hence the higher the quality of encryption. So, the quality of encryption may be expressed by several metrics, such as histogram uniformity, histogram deviation, and correlation coefficient between original and encrypted images.

4.3.1 Histogram Analysis

Figure 4.5 shows the histogram results of the three above-mentioned tested simulation cases. It is observed that the full proposed hybrid encryption framework with Rubik's cube encryption technique gives approximately-flat histograms for all simulation tests, which means high and good quality of encryption. Therefore, the obtained results of the tested simulation cases prove the importance of the utilization of the Rubik's cube encryption technique incorporated with the proposed hybrid encryption techniques compared to the case of not employing any encryption techniques for all tested image streams.

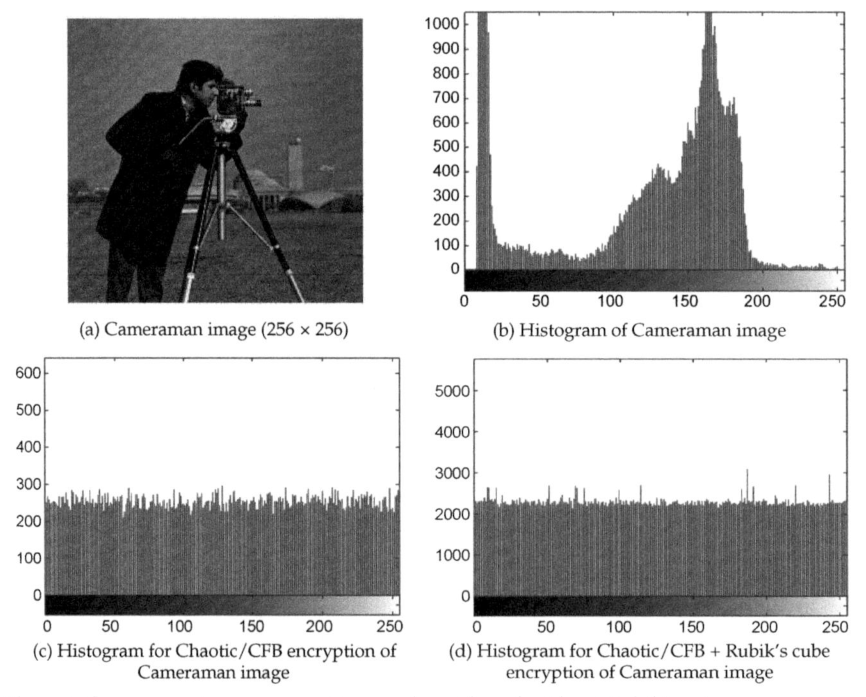

(a) Cameraman image (256 × 256)

(b) Histogram of Cameraman image

(c) Histogram for Chaotic/CFB encryption of Cameraman image

(d) Histogram for Chaotic/CFB + Rubik's cube encryption of Cameraman image

Fig. 4.5: Cameraman image encryption results with and without Rubik's cube.

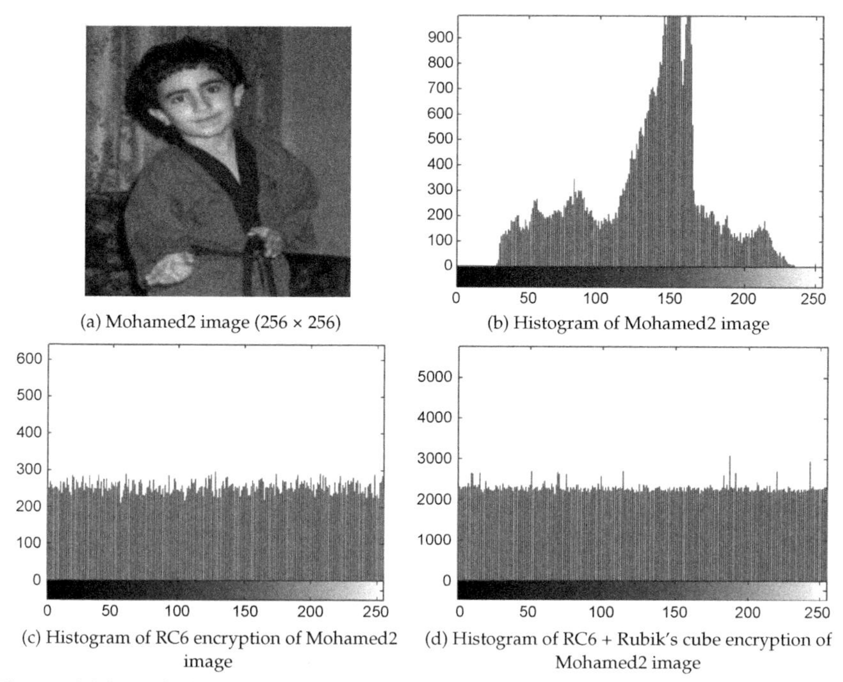

(a) Mohamed2 image (256 × 256)

(b) Histogram of Mohamed2 image

(c) Histogram of RC6 encryption of Mohamed2 image

(d) Histogram of RC6 + Rubik's cube encryption of Mohamed2 image

Fig. 4.6: Mohamed2 image encryption results with and without Rubik's cube.

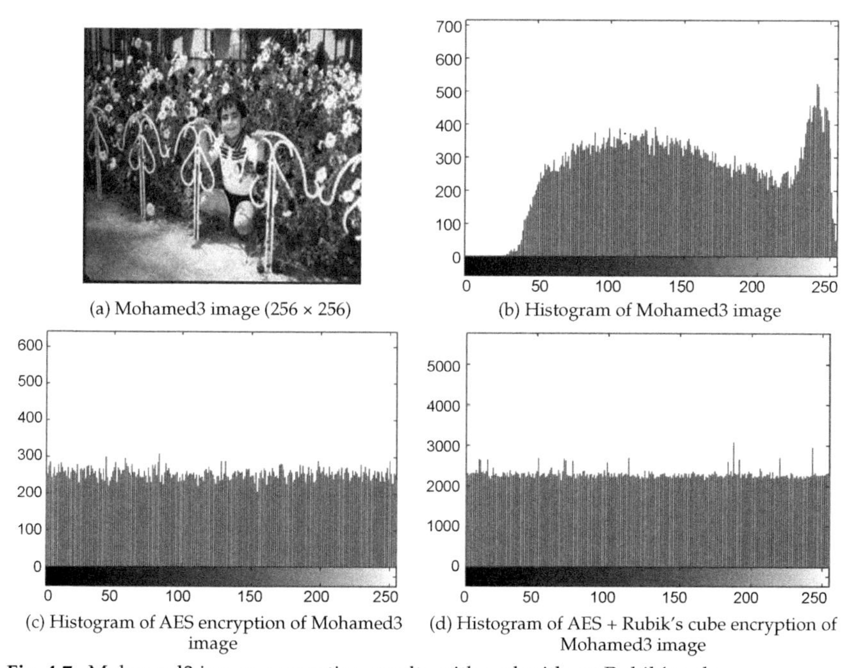

(a) Mohamed3 image (256 × 256)

(b) Histogram of Mohamed3 image

(c) Histogram of AES encryption of Mohamed3 image

(d) Histogram of AES + Rubik's cube encryption of Mohamed3 image

Fig. 4.7: Mohamed3 image encryption results with and without Rubik's cube.

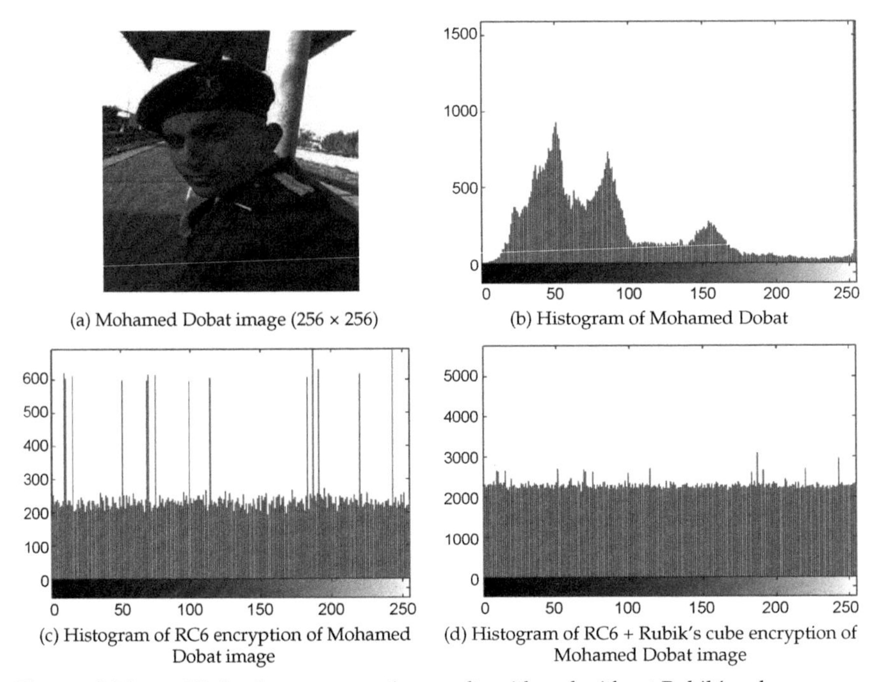

(a) Mohamed Dobat image (256 × 256)

(b) Histogram of Mohamed Dobat

(c) Histogram of RC6 encryption of Mohamed Dobat image

(d) Histogram of RC6 + Rubik's cube encryption of Mohamed Dobat image

Fig. 4.8: Mohamed Dobat image encryption results with and without Rubik's cube.

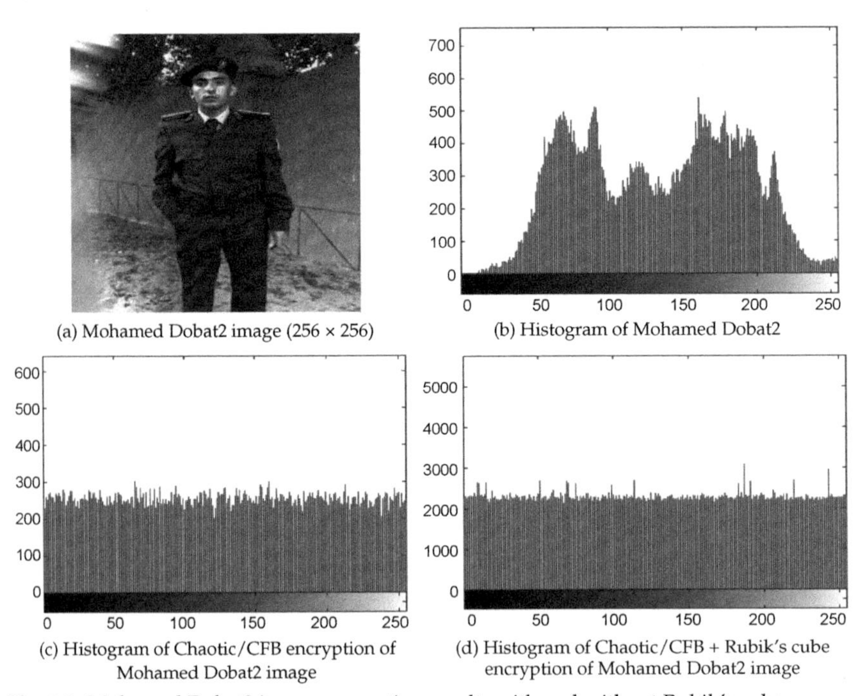

(a) Mohamed Dobat2 image (256 × 256)

(b) Histogram of Mohamed Dobat2

(c) Histogram of Chaotic/CFB encryption of Mohamed Dobat2 image

(d) Histogram of Chaotic/CFB + Rubik's cube encryption of Mohamed Dobat2 image

Fig. 4.9: Mohamed Dobat2 image encryption results with and without Rubik's cube.

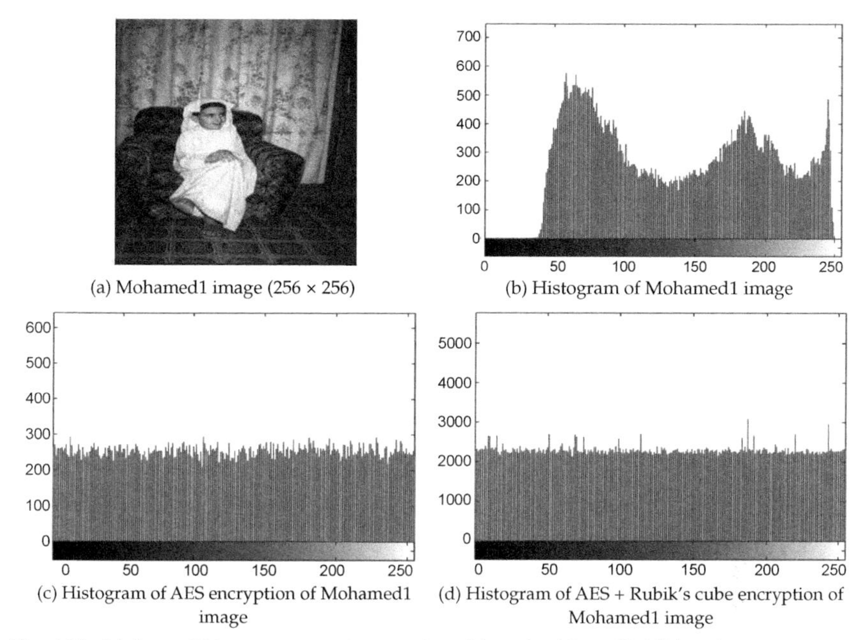

(a) Mohamed1 image (256 × 256)

(b) Histogram of Mohamed1 image

(c) Histogram of AES encryption of Mohamed1 image

(d) Histogram of AES + Rubik's cube encryption of Mohamed1 image

Fig. 4.10: Mohamed1 image encryption results with and without Rubik's cube.

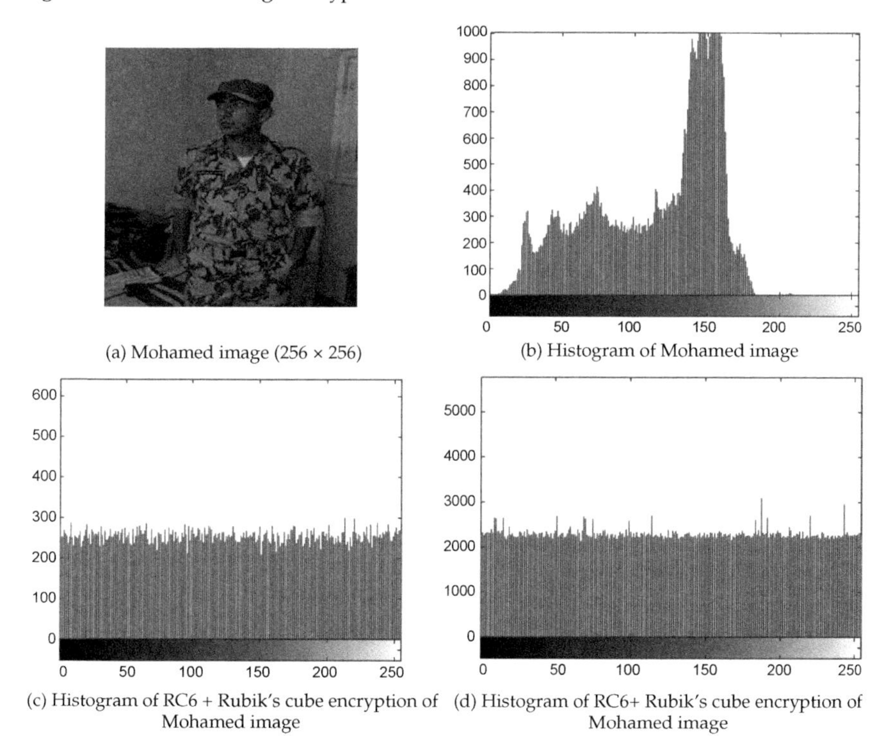

(a) Mohamed image (256 × 256)

(b) Histogram of Mohamed image

(c) Histogram of RC6 + Rubik's cube encryption of Mohamed image

(d) Histogram of RC6+ Rubik's cube encryption of Mohamed image

Fig. 4.11: Mohamed image encryption results with and without Rubik's cube.

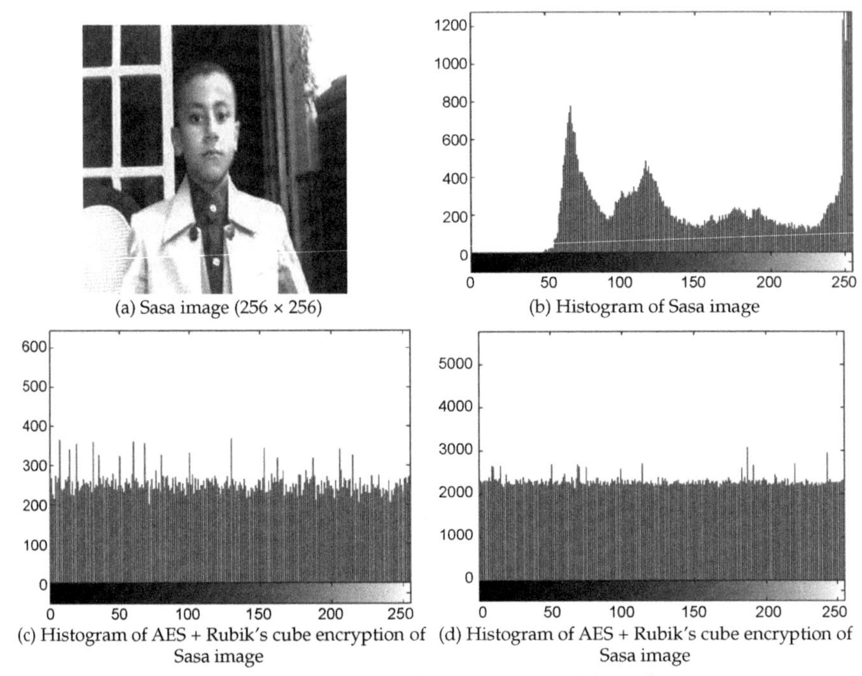

(a) Sasa image (256 × 256)

(b) Histogram of Sasa image

(c) Histogram of AES + Rubik's cube encryption of Sasa image

(d) Histogram of AES + Rubik's cube encryption of Sasa image

Fig. 4.12: Sasa image encryption results with and without Rubik's cube.

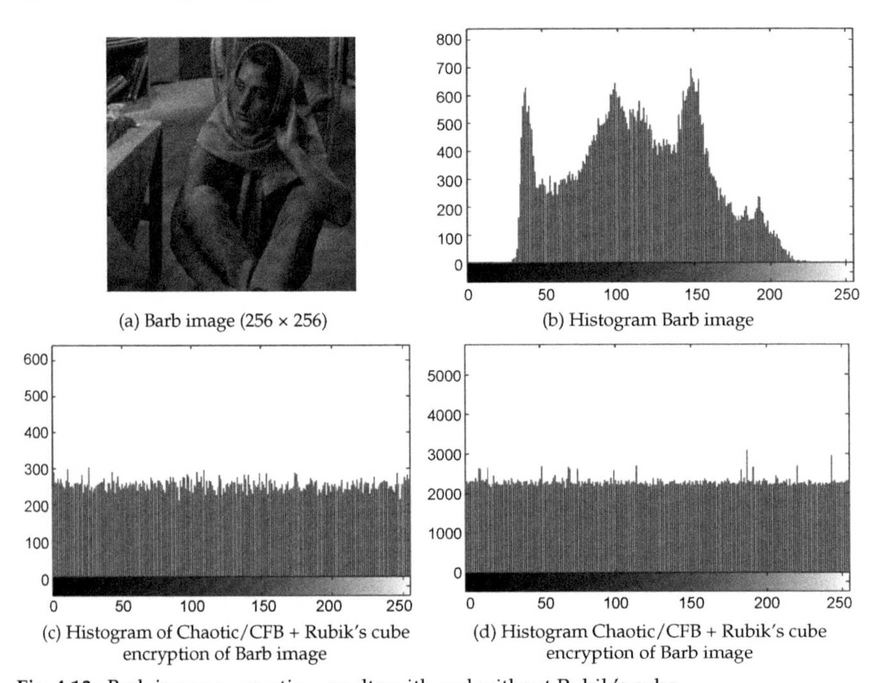

(a) Barb image (256 × 256)

(b) Histogram Barb image

(c) Histogram of Chaotic/CFB + Rubik's cube encryption of Barb image

(d) Histogram Chaotic/CFB + Rubik's cube encryption of Barb image

Fig. 4.13: Barb image encryption results with and without Rubik's cube.

4.3.2 Deviation and Correlation Coefficient

Tables 4.1 and 4.2 show the average deviation and correlation comparison values of the nine different images of the jointly encrypted Cameraman, Mohamed2, Mohamed3, Mohamed Dobat, Mohamed Dobat2, Mohamed1, Mohamed, Sasa, and Barb images with the hybrid encryption technique with and without using the proposed Rubik's cube encryption technique

Table 4.1: Deviation values for the jointly encrypted image Cameraman, Mohamed2, Mohamed3, Mohamed Dobat, Mohamed Dobat2, Mohamed1, Mohamed, Sasa, and Barb 2-D images.

Encryption Technique	Deviation
AES encryption of Cameraman	35.18
ASE + Rubik's cube encryption of Cameraman (Similar faces)	35.25
AES encryption of Mohamed	29.73
ASE + Rubik's cube encryption of Mohamed (Similar faces)	35.36
RC6 encryption of Cameraman	35.11
RC6 + Rubik's cube encryption of Cameraman (Similar faces)	35.61
RC6 encryption of Mohamed	29.66
RC6 + Rubik's cube encryption of Mohamed (Similar faces)	30.03
RC6 encryption of Lena	34.39
RC6 + Rubik's cube encryption of Lena (Similar faces)	34.67
Chaotic/CBC encryption of Mohamed (Similar faces)	35.24
Chaotic/CBC + Rubik's cube encryption of Mohamed (Similar faces)	35.52
Chaotic/CFB encryption of Mohamed (Similar faces)	35.63
Chaotic/CFB + Rubik's cube encryption of Mohamed (Similar faces)	35.72
Chaotic/OFB encryption of Mohamed (Similar faces)	35.63
Chaotic/OFB + Rubik's cube encryption of Mohamed (Similar faces)	35.74
Chaotic/CFB + Rubik's cube encryption of cameraman (Different faces)	35.63
RC6 + Rubik's cube encryption of Mohamed2 (Different faces)	38.22
AES + Rubik's cube encryption of Mohamed3 (Different faces)	50.66
RC6 + Rubik's cube encryption of Mohamed Dobat (Different faces)	34.71
Chaotic/CFB + Rubik's cube encryption of Mohamed Dobat2 (Different faces)	41.51
AES + Rubik's cube encryption of Mohamed1 (Different faces)	44.81
RC6 + Rubik's cube encryption of Mohamed (Different faces)	30.03
AES + Rubik's cube encryption of Sasa (Different faces)	64.86
Chaotic/CFB + Rubik's cube encryption of Barb (Different faces)	28.71
Average value of Different images with different algorithm + Rubik's cube algorithm (Different faces)	41.02

Table 4.2: Correlation values for the jointly encrypted image Cameraman, Mohamed2, Mohamed3, Mohamed Dobat, Mohamed Dobat2, Mohamed1, Mohamed, Sasa, Barb 2-D images.

Encryption Technique	Correlation
AES encryption of Cameraman	0.0077
ASE + Rubik's cube encryption of Cameraman (Similar faces)	–0.0034
AES encryption of Mohamed	–4.2268e-04
ASE + Rubik's cube encryption of Mohamed (Similar faces)	–0.0067
RC6 encryption of Cameraman	0.0023
RC6 + Rubik's cube encryption of Cameraman (Similar faces)	–0.0016
RC6 encryption of Mohamed	0.0042
RC6 + Rubik's cube encryption of Mohamed (Similar faces)	0.0026
RC6 encryption of Lena	–8.6183e-04
RC6 + Rubik's cube encryption of Lena (Similar faces)	0.0043
Chaotic/CBC encryption of Mohamed (Similar faces)	0.0050
Chaotic/CBC + Rubik's cube encryption of Mohamed (Similar faces)	–0.0038
Chaotic/CFB encryption of Mohamed (Similar faces)	0.0015–
Chaotic/CFB + Rubik's cube encryption of Mohamed (Similar faces)	–0.0067
Chaotic/OFB encryption of Mohamed (Similar faces)	0.0050–
Chaotic/OFB + Rubik's cube encryption of Mohamed (Similar faces)	–0.0033
Chaotic/CFB + Rubik's cube encryption of cameraman (Different faces)	–0.0071
RC6 + Rubik's cube encryption of Mohamed2 (Different faces)	3.1934e-04
AES + Rubik's cube encryption of Mohamed3 (Different faces)	–0.0071
RC6 + Rubik's cube encryption of Mohamed Dobat (Different faces)	–3.5834e-04
Chaotic/CFB + Rubik's cube encryption of Mohamed Dobat2 (Different faces)	–0.0045
AES + Rubik's cube encryption of Mohamed1 (Different faces)	–0.0034
RC6 + Rubik's cube encryption of Mohamed (Different faces)	0.0014
AES + Rubik's cube encryption of Sasa (Different faces)	0.0052
Chaotic/CFB + Rubik's cube encryption of Barb (Different faces)	0.0053
Average value of Different images with different algorithm + Rubik's cube algorithm (Different faces)	–0.0011

it's noticed that the proposed hybrid technique is recommended for 3-D encryption as compared to the traditional techniques [98–101]. Therefore, the obtained results of the tested simulation cases prove the importance of the utilization of the Rubik's cube encryption technique incorporated with the proposed hybrid encryption technique compared to the case of not employing any encryption technique for all tested 3-D streams.

4.3.3 Processing Time

Table 4.3 shows the average time of calculations of the different tested cases of the proposed encryption techniques with and without employing the Rubik's cube encryption algorithm. It is noticed that the case of using Rubik's cube encryption algorithm needs less time in processing, as shown in Table 4.3. The real time of the Rubik's cube itself is very short. The

Table 4.3: Average processing times image for the jointly encrypted images with the proposed hybrid encryption techniques with and without Rubik's cube.

AES encryption of Cameraman	122.57 seconds
ASE + Rubik's cube encryption of Cameraman (Similar faces)	725.17 seconds
AES encryption of Mohamed	112.65 seconds
ASE + Rubik's cube encryption of Mohamed (Similar faces)	830.06 seconds
RC6 encryption of Cameraman	653.20 seconds
RC6 + Rubik's cube encryption of Cameraman (Similar faces)	9806.57 seconds
RC6 encryption of Mohamed	724.41 seconds
RC6 + Rubik's cube encryption of Mohamed (Similar faces)	5567.08 seconds
RC6 encryption of Lena	1243.65 seconds
RC6 + Rubik's cube encryption of Lena (Similar faces)	9109.56 seconds
Chaotic/CBC encryption of Mohamed (Similar faces)	00.80 seconds
Chaotic/CBC + Rubik's cube encryption of Mohamed (Similar faces)	10.03 seconds
Chaotic/CFB encryption of Mohamed (Similar faces)	01.06 seconds
Chaotic/CFB + Rubik's cube encryption of Mohamed (Similar faces)	10.06 seconds
Chaotic/OFB encryption of Mohamed (Similar faces)	3.64 seconds
Chaotic/OFB + Rubik's cube encryption of Mohamed (Similar faces)	11.73 seconds
Chaotic/CFB + Rubik's cube encryption of cameraman (Different faces)	2.90 seconds
RC6 + Rubik's cube encryption of Mohamed2 (Different faces)	1491.73 seconds
AES + Rubik's cube encryption of Mohamed3 (Different faces)	1914.02 seconds
RC6 + Rubik's cube encryption of Mohamed Dobat (Different faces)	960.72 seconds
Chaotic/CFB + Rubik's cube encryption of Mohamed Dobat2 (Different faces)	2.27 seconds
AES + Rubik's cube encryption of Mohamed1 (Different faces)	1901.82 seconds
RC6 + Rubik's cube encryption of Mohamed (Different faces)	1593.17 seconds
AES + Rubik's cube encryption of Sasa (Different faces)	777.61 seconds
Chaotic/CFB + Rubik's cube encryption of Barb (Different faces)	2.70 seconds
Average value of Different images with different algorithm + Rubik's cube algorithm (Different faces)	960.55 seconds

computational time of the traditional RC6, AES, and chaotic encryption techniques can be reduced significantly with parallel processing.

4.4 Simulation Results

To evaluate the performance of the suggested hybrid encryption framework, several simulation tests have been carried out on standard images (Cameraman, Mohamed2, Mohamed3, Mohamed Dobat, Mohamed Dobat2, Mohamed1, Mohamed, Sasa, and Barb) stream [102]. The tested sequences have different spatial and temporal characteristics. In the simulation results, after encryption stage using different AES, chaotic encryption with CFB mode, and RC6 encryption algorithms with and without Rubik's cube technique, the encrypted images are digitized and transmitted via the OFDM communication system through AWGN and Rayleigh fading channels. In all the simulation tests, OFDM is preceded by QPSK modulation with 128 sub-channels, each with 500 kHz spacing. The summarized simulation parameters are mentioned in Table 4.4.

To clarify the influence of employing the suggested hybrid encryption framework, we tested it with previous different simulation cases [98–101]. The tested simulation case is the selection of the nine images to be the faces of the proposed Rubik's cube, as shown in Figs. 4.14 and 4.15. Then, we encrypt each one of these nine images on the Rubik's cube faces individually with different encryption techniques, including chaotic encryption with CFB mode, RC6, and AES. We have presented simulation results to check the best encryption technique to be selected for each image of the Rubik's cube faces. Finally, we find that the best encryption results can be achieved in the case of choosing the key of encryption techniques for the nine images of the Rubik's cube faces as [chaotic with CFB mode,

Table 4.4: Simulation parameters.

System bandwidth	64 MHz
Modulation type	QPSK
Cyclic prefix length	16 samples
Number of sub-carriers	128
Sub-carriers spacing	500 kHz
Channel fading type	Rayleigh fading channel
Noise environment	AWGN
Equalization type	LMMSE
Type of transform	FFT, DWT, and DCT
Rubik's cube and chaotic keys	IVs of each image
RC6 key	$(00010203040506070809)_{16}$

(a) Cameraman image (b) Mohamed2 image (c) Mohamed3 image
(d) Mohamed Dobat image (e) Mohamed Dobat2 image (f) Mohamed1 image
(g) Mohamed1 image (h) Sasa image (i) Barb image

Fig. 4.14: Original samples of different images.

Fig. 4.15: Converted 3-D Rubik's cube faces into a 2-D image of the images in Fig. (4.14).

RC6, AES, RC6, chaotic with CFB chaotic, AES, RC6, AES, chaotic with CFB mode].

We compared the performance of the proposed encryption techniques with and without employing the proposed Rubik's cube encryption.

Tables 4.2 and 4.3 show the deviation and correlation comparison values of the nine images of the 3-D sequences of the proposed hybrid encryption techniques with and without using the proposed Rubik's cube encryption.

Tables 4.5 and 4.6 show the average PSNR performance vs. different values of E_b/N_o for the OFDM system with its three different versions over the Rayleigh fading channel at f_d = 600 Hz in the case of nine different images on the Rubik's cube faces to be jointly transmitted with and without employing the proposed Rubik's cube technique. From all obtained average PSNR values of the three tested simulation cases, the results prove the significance of exploiting the proposed Rubik's cube encryption technique combined with the suggested hybrid encryption framework for achieving better performance compared to the utilization of the traditional methods [102, 103]. Also, it is noticed that the average PSNR results of the FFT-OFDM system are better than those of the DWT and DCT systems for all tested images. Figures 4.16 to 4.24 illustrate the subjective comparison results of the jointly encrypted Cameraman, Mohamed2, Mohamed3, Mohamed Dobat, Mohamed Dobat2, Mohamed1, Mohamed, Sasa, Barb 2-D images for transmission through the OFDM system with its three different versions. Transmission is performed over the Rayleigh fading channel at f_d = 600 Hz at E_b/N_o = 2, 6, and 10 dB.

It is observed from the subjective comparison results that the performance of the FFT-OFDM system is better than those of the DCT and DWT systems for all tested images, where the decrypted images are received typically similar to the original images with high quality. Therefore, from all presented objective and subjective results, it is recognized that the proposed hybrid encryption techniques with the Rubik's cube encryption are more appreciated and recommended for image transmission compared to the traditional techniques [103]. The full proposed hybrid encryption framework has preferable objective and visual simulation results compared to the case of not exploiting the proposed Rubik's cube technique [103]. Furthermore, it is noticed that the suggested hybrid encryption framework offers adequate experimental results for various standard images that have different spatial and temporal features.

Finally, this chapter presented an improved hybrid encryption framework for efficient image transmission over wireless channels. The major contribution of this work is the inclusion of Rubik's cube encryption technique with a hybrid structure of AES, RC6, and chaotic encryption techniques. So, the proposed hybrid encryption framework adds a better degree of both permutation and diffusion to the encrypted images, simultaneously. Experimental simulation results verified the heartening achievement of the proposed hybrid encryption framework in efficiently encrypting the transmitted image sequences. Therefore, it is more qualified

Table 4.5: Average PSNR performance vs. different values of E_b/N_o for the jointly encrypted images with the proposed hybrid encryption techniques **with Rubik's cube** technique of the different 2-D images transmission via FFT/DCT/DWT-OFDM system over Rayleigh fading channel at $f_d = 600$ Hz.

Jointly Encrypted Images	FFT-OFDM (E_b/N_o = 2 dB)	FFT-OFDM (E_b/N_o = 6 dB)	FFT-OFDM (E_b/N_o = 10 dB)	DCT-OFDM (E_b/N_o = 2 dB)	DCT-OFDM (E_b/N_o = 6 dB)	DCT-OFDM (E_b/N_o = 10 dB)	DWT-OFDM (E_b/N_o = 2 dB)	DWT-OFDM (E_b/N_o = 6 dB)	DWT-OFDM (E_b/N_o = 10 dB)
Cameraman encrypted image with Chaotic/CFB	39.71	Inf.	Inf.	32.60	43.39	Inf.	33.39	43.39	Inf.
Mohamed2 encrypted image with RC6	30.76	45.15	Inf.	28.29	32.85	37.02	26.86	31.93	Inf.
Mohamed 3 encrypted image with AES	30.10	45.15	Inf.	27.67	34.01	36.12	26.79	31.54	Inf.
Mohamed Dobat encrypted image with RC6	32.60	48.16	Inf.	30.68	35.86	39.71	30.24	35.39	48.1648
Mohamed Dobat2 encrypted image with Chaotic/CFB	40.38	Inf.	Inf.	35.61	43.39	45.15	35.15	45.15	Inf.
Mohamed1 encrypted image with AES	30.53	Inf.	Inf.	28.34	33.39	39.71	27.30	33.5408	43.3936
Mohamed encrypted image with RC6	30.45	45.15	Inf.	27.91	32.14	39.13	27.16	32.9797	45.1545
Sasa encrypted image with AES	29.29	Inf.	Inf.	27.99	34.74	37.02	26.99	33.1133	42.1442
Barb encrypted image with Chaotic/CFB	38.62	48.16	Inf.	32.85	41.17	Inf.	34.02	48.1648	Inf.

Table 4.6: Average PSNR performance vs. different values of E_b/N_o for the jointly encrypted images with the proposed hybrid encryption techniques **without Rubik's cube** technique of the different 2-D images transmission via FFT/DCT/DWT-OFDM system over Rayleigh fading channel at $f_d = 600$ Hz.

Jointly Encrypted Images	FFT-OFDM (E_b/N_o = 2 dB)	FFT-OFDM (E_b/N_o = 6 dB)	FFT-OFDM (E_b/N_o = 10 dB)	DCT-OFDM (E_b/N_o = 2 dB)	DCT-OFDM (E_b/N_o = 6 dB)	DCT-OFDM (E_b/N_o = 10 dB)	DWT-OFDM (E_b/N_o = 2 dB)	DWT-OFDM (E_b/N_o = 6 dB)	DWT-OFDM (E_b/N_o = 10 dB)
Cameraman encrypted image with Chaotic/CFB	23.62	40.93	Inf.	16.70	26.18	31.36	16.55	29.21	43.12
Mohamed2 encrypted image with RC6	13.75	33.83	46.44	11.04	16.39	21.75	10.62	17.83	31.47
Mohamed 3 encrypted image with AES	12.68	32.59	46.78	9.81	15.40	20.62	9.49	16.57	28.84
Mohamed Dobat encrypted image with RC6	11.8971	37.4047	Inf.	9.00	14.51	19.74	8.59	15.77	29.33
Mohamed Dobat2 encrypted image with Chaotic/CFB	23.50	44.83	96.29	16.59	26.62	32.24	16.97	29.21	41.59
Mohamed1 encrypted image with AES	12.61	33.62	43.13	10.04	15.37	20.56	9.54	16.35	31.29
Mohamed encrypted image with RC6	13.69	33.15	41.22	11.07	16.55	21.62	10.47	17.78	30.97
Sasa encrypted image with AES	11.69	33.74	43.93	8.88	14.46	20.12	8.36	15.97	27.80
Barb encrypted image with Chaotic/CFB	23.96	50.17	96.29	16.73	26.26	32.20	16.87	28.79	43.82

Fig. 4.16: Subjective comparison results of the jointly encrypted Cameraman, Mohamed2, Mohamed3, Mohamed Dobat, Mohamed Dobat2, Mohamed1, Mohamed, Sasa, Barb images for transmission through the DCT-OFDM system over Rayleigh fading channel at $f_d = 600$ Hz and $E_b/N_o = 2$ dB with the proposed hybrid encryption techniques implementing Rubik's cube encryption.

Fig. 4.17: Subjective comparison results of the jointly encrypted Cameraman, Mohamed2, Mohamed3, Mohamed Dobat, Mohamed Dobat2, Mohamed1, Mohamed, Sasa, Barb images for transmission through the DWT-OFDM system over Rayleigh fading channel at $f_d = 600$ Hz and $E_b/N_o = 2$ dB with the proposed hybrid encryption techniques implementing Rubik's cube encryption.

Fig. 4.18: Subjective comparison results of the jointly encrypted Cameraman, Mohamed2, Mohamed3, Mohamed Dobat, Mohamed Dobat2, Mohamed1, Mohamed, Sasa, Barb images for transmission through the FFT-OFDM system over Rayleigh fading channel at $f_d = 600$ Hz and $E_b/N_o = 2$ dB with the proposed hybrid encryption techniques implementing Rubik's cube encryption.

Fig. 4.19: Subjective comparison results of the jointly encrypted Cameraman, Mohamed2, Mohamed3, Mohamed Dobat, Mohamed Dobat2, Mohamed1, Mohamed, Sasa, Barb images for transmission through the DCT-OFDM system over Rayleigh fading channel at $f_d = 600$ Hz and $E_b/N_o = 6$ dB with the proposed hybrid encryption techniques implementing Rubik's cube encryption.

Fig. 4.20: Subjective comparison results of the jointly encrypted Cameraman, Mohamed2, Mohamed3, Mohamed Dobat, Mohamed Dobat2, Mohamed1, Mohamed, Sasa, Barb images for transmission through the DWT-OFDM system over Rayleigh fading channel at $f_d = 600$ Hz and $E_b/N_o = 6$ dB with the proposed hybrid encryption techniques implementing Rubik's cube encryption.

Fig. 4.21: Subjective comparison results of the jointly encrypted Cameraman, Mohamed2, Mohamed3, Mohamed Dobat, Mohamed Dobat2, Mohamed1, Mohamed, Sasa, Barb images for transmission through the FFT-OFDM system over Rayleigh fading channel at $f_d = 600$ Hz and $E_b/N_o = 6$ dB with the proposed hybrid encryption techniques implementing Rubik's cube encryption.

Fig. 4.22: Subjective comparison results of the jointly encrypted Cameraman, Mohamed2, Mohamed3, Mohamed Dobat, Mohamed Dobat2, Mohamed1, Mohamed, Sasa, Barb images for transmission through the DCT-OFDM system over Rayleigh fading channel at $f_d = 600$ Hz and $E_b/N_o = 10$ dB with the proposed hybrid encryption techniques implementing Rubik's cube encryption.

Fig. 4.23: Subjective comparison results of the jointly encrypted Cameraman, Mohamed2, Mohamed3, Mohamed Dobat, Mohamed Dobat2, Mohamed1, Mohamed, Sasa, Barb images for transmission through the DWT-OFDM system over Rayleigh fading channel at $f_d = 600$ Hz and $E_b/N_o = 10$ dB with the proposed hybrid encryption techniques implementing Rubik's cube encryption.

Fig. 4.24: Subjective comparison results of the jointly encrypted Cameraman, Mohamed2, Mohamed3, Mohamed Dobat, Mohamed Dobat2, Mohamed1, Mohamed, Sasa, Barb images for transmission through the FFT-OFDM system over Rayleigh fading channel at $f_d = 600$ Hz and $E_b/N_o = 10$ dB with the proposed hybrid encryption techniques implementing Rubik's cube encryption.

for secure image transmission. It provides appreciated deviation, correlation, and histogram results. The proposed hybrid encryption framework has proved its ability to adequately encrypt various standard image streams that have different temporal and spatial characteristics with peak image quality. Moreover, the simulation results expounded the prominence of deploying the hybrid structure of the proposed Rubik's cube encryption technique with the traditional encryption techniques in order to reinforce the subjective image subjective quality and likewise acquire considerable objective results.

Chapter 5

Proposed Hybrid Encryption Framework for Reliable 3-D Wireless Video Communications

5.1 Introduction

In this chapter, a 3-D Video (3DV) hybrid encryption framework based on the Rubik's cube is suggested as a means to attain simultaneous encryption of a group of 3DV frames. After that, the obtained encrypted 3DV frames are further encrypted in a second stage with Rubik's cube technique. Chaotic, RC6, or AES encrypted 3DV frames are utilized as the faces of the Rubik's cube.

The standard of Multi-view Video Compression (MVC) [104, 105] attains efficient 3-D video encoding. It has obtained a lot of attention recently. The MVC standard is a complement of the 2-D video compression standard [106], and it is predicted to quickly take the place of the 2-D videos in various implementations like education, medicine, 3DTV, entertainment, and gaming. In the 3-D-MVC system, the 3DV is composed of diverse sequences picked for the selfsame object with different cameras. Therefore, it is an important mission to attain enough compression to fulfill future bandwidth requirements, whilst preserving a decisive 3DV reception quality. The 3DV transmission over wireless networks has increased dramatically [107, 108]. To transmit 3DV over limited-resources channels, a strong enough encoding process must be utilized, whilst maintaining an appreciated 3DV reception quality. For efficient 3DV communication, the 3-D-MVC system must exploit the spatial and temporal matching among neighboring frames in the selfsame video in addition to the inter-view correlation inside various 3-D video sequences in order to improve the compression efficiency. However, video compression with high rates is more susceptible to transmission channel corruptions.

The 3DV transmission through wireless channels is permanently subject to packet errors of both burst and random natures [109, 110]. The predictive 3-D MVC coding framework introduced in Fig. 5.1 [111, 112] is used to compress the transported 3DV streams. It consists of the inter P and B encoded frames and the intra I encoded frames. It is known that the inter-frames in the odd B views are encoded via the intra-frames in the I view, and through the inter-frames within the even P views. Therefore, the faults might propagate to the next frames or to the contiguous views, and thus create destitute 3DV quality.

The demand for high speed and secure wireless 3-D multimedia communication systems has grown in the last few decades. The progress in mobile communications, satellite applications, Internet applications, and computer networks has given rise to new problems regarding security and privacy [113–115]. Having a secure and reliable means for communicating with images and video is becoming a necessity, and its related issues must be carefully considered. Hence, network security and data encryption have become important [84–85]. Image and video encryption have applications in various fields, including Internet communication, multimedia systems, medical imaging, telemedicine, and military communications [117–118]. Therefore, due to the fast progress in network development, humans can

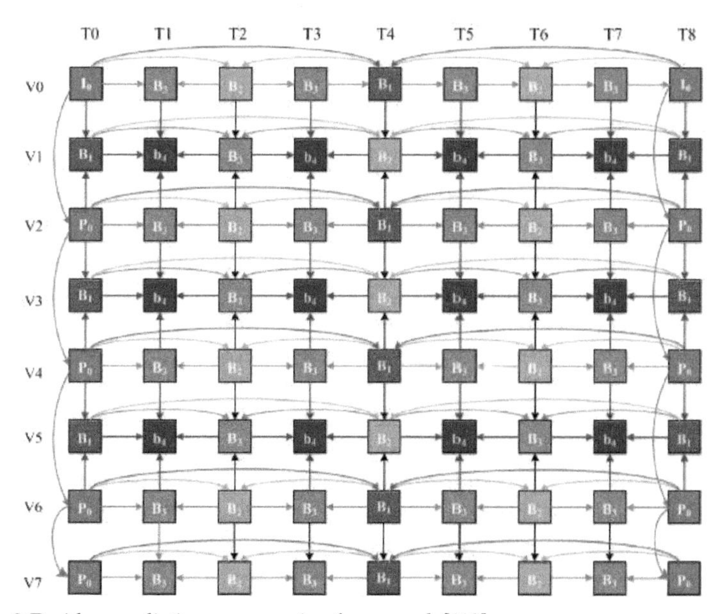

Fig. 5.1: 3-D video prediction compression framework [112].

easily and arbitrarily distribute or access digital multimedia data from networks. The ownership security has become an important issue for individuals, and it requires more attention. Thus, there is a significant threat to copyright owners and digital multimedia producers to conserve multimedia from intruder prospection to avoid losses in transmitted data [86].

Encryption is one of the most favorable methods to secure digital multimedia files in the domains of copyright protection and data authentication. There are two main types of applications for image and video encryption in wireless communication systems: online applications and web page applications [87]. In online applications, such as video encryption, the speed of the encryption algorithm and the immunity to wireless communication channel degradations are of major concern. On the other hand, web page applications are mainly concerned with the security issues. The speed of encryption is not the major factor, and noise-free channels can be assumed [88].

The objective of this chapter is to introduce an efficient 3DV hybrid encryption framework from the security perspective for the 3-D frames to be transmitted over a wireless communication channel, while maintaining the good quality of the received 3-D frames. We have adopted OFDM as the communication system for wireless transmission of the encrypted 3DV frames. Different versions of OFDM, such as FFT-OFDM, DCT, and DWT, are studied in this chapter in order to select the most appropriate modulation scheme that can guarantee our objectives of high security of encryption and good quality of received 3DV frames. Various channel impairments, such as AWGN and fading, are considered in this chapter. Moreover, channel equalization is investigated in order to study its effect on the quality of the received 3DV frames. So, we concentrate in our study on the ability of the suggested 3DV hybrid encryption framework to deal with channel degradations. In the proposed hybrid encryption framework, we need to achieve both diffusion and permutation in the encrypted 3DV frames. To this aim, we can use chaotic, RC6, or AES technique in a pre-processing step to achieve the permutation and diffusion. The Rubik's cube is used afterwards to achieve a larger degree of permutation.

The Rubik's cube technique will be used to encrypt a group of 3DV frames, simultaneously. The 3-D frames implemented on these nine faces are those encrypted firstly with the chaotic, RC6, or AES technique. The proposed hybrid encryption framework will guarantee both diffusion and permutation in the encrypted 3DV frames. We will study its sensitivity to the wireless channel impairments, and also study the effect of channel equalization on the received 3DV frames quality. Figure 5.2 illustrates the Rubik's cube with nine frames.

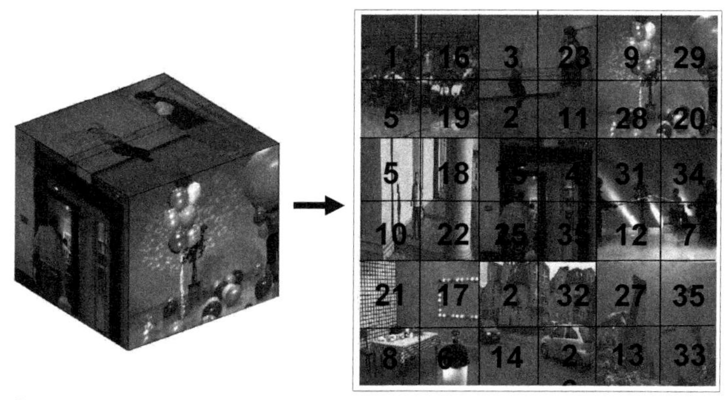

Fig. 5.2: Converting 3-D Rubik's cube with frames into a 2-D image.

5.2 Proposed Hybrid Encryption Framework

In this section, the suggested hybrid encryption framework for efficient 3-D video transmission over OFDM is introduced. The general structure of the proposed 3DV communication system is shown in Fig. 5.3. First, the huge amount of data of the 3DV stream is compressed utilizing the 3-D H.264/MVC encoder to reduce the 3DV data size to be compatible with the transmission requirements over limited-resources wireless channels. After that, the compressed 3DV bit streams are encrypted using the proposed hybrid encryption framework and transmitted through OFDM system. Then, the 3DV data are received, decrypted, and decoded by the 3-D H.264/MVC decoder.

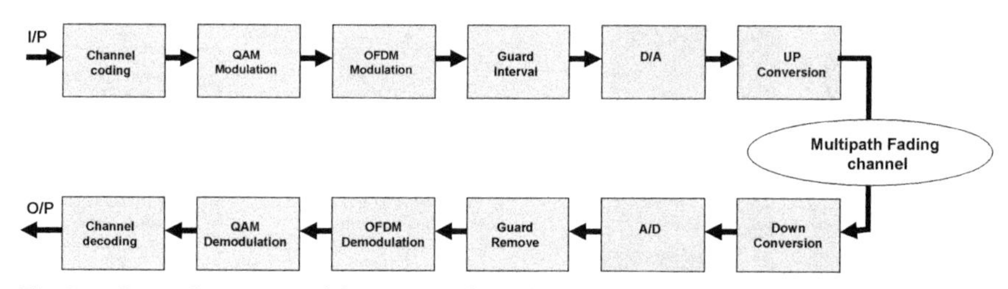

Fig. 5.3: General structure of the proposed wireless 3-D video communication system.

Figure 5.3 shows the key steps in the OFDM communication system, which is used for the transmission of the encrypted 3DV frames. The first step is 3DV frame encryption. Then, the encrypted 3DV frames are transformed over the OFDM system. In the proposed 3DV communication system, the 3-D video frames to be transmitted are first encrypted by the proposed hybrid encryption framework, as illustrated in Fig. 5.4. The encryption and communication processes can be summarized as follows:

Input the database of images of the database to encrypt;
- initialize all process;
- receive the image bit-streams;
- Start with chaotic, AES, and RC6 encryption algorithm in a defined key;

for all Encrypted images, ***do***
- Set the encrypted images over Rubik's cube nine faces;
- Convert the whole faces into a 2-D image with larger size;
- Randomize the new image with Rubik's cube mechanism;
- Transform the image into binary format;

end for

if Proposed Hybrid Encryption algorithm is ***done, then***
- Perform OFDM modulation;
- Transmit the signal over the wireless channel;
- Perform channel equalization at the receiver;
- Transform the received signal into binary format;
- Reconstruct the image into pixel values;

end if

for all Received encrypted images, ***do***
- Return to the canonical basis to get the decrypted images;

end for

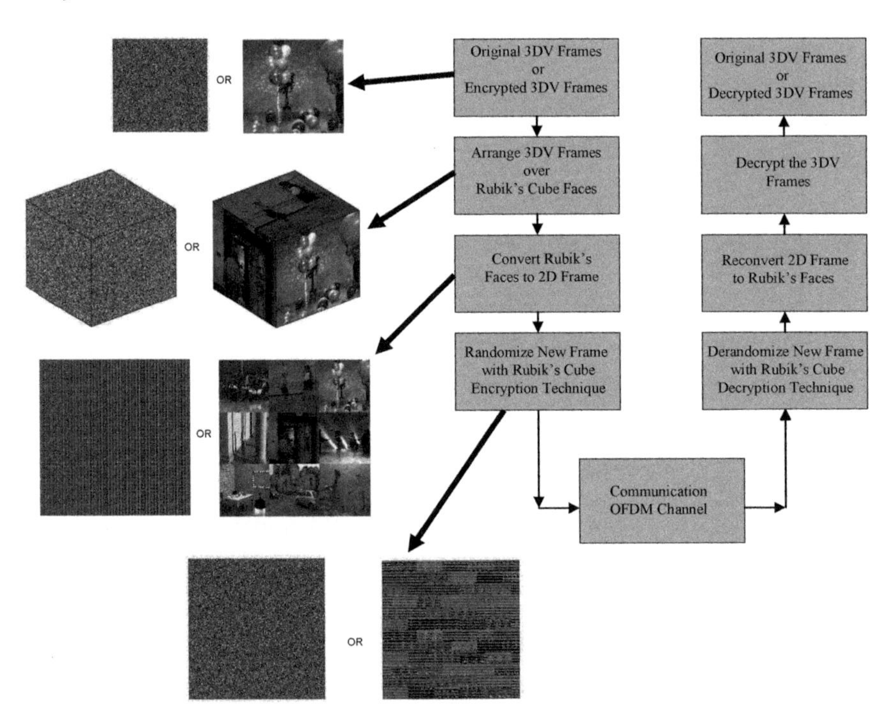

Fig. 5.4: The proposed hybrid encryption and communication framework.

5.3 Encryption Quality Evaluation Metrics

Both histogram uniformity, histogram deviation, and correlation coefficient between the original and encrypted frames are considered for evaluation. To clarify the influence of employing the suggested hybrid encryption framework, we tested three different simulation cases. The first tested simulation case is the selection of the first nine frames of the Poznan Street and Shark 3DV streams separately to be the faces of the proposed Rubik's cube, as shown in Figs. 5.5 and 5.6. Then, we encrypt each one of these nine frames of the Rubik's cube faces of each 3DV stream individually with different encryption techniques from the proposed chaotic with CFB mode, RC6, and AES algorithms. We have carried out a lot of simulation results to check the best encryption technique that must be selected for each frame of the Rubik's cube faces which achieve the best encryption performance. Finally, we find that the best encryption results can be achieved in the case of choosing the key of encryption techniques for the nine frames of the Rubik's cube faces as [chaotic with CFB mode, RC6, AES, RC6, chaotic with CFB mode, AES, RC6, AES, chaotic with CFB mode], respectively. We compare the performance of the proposed encryption techniques with and without employing the proposed Rubik's cube encryption technique.

The second tested simulation case is the selection of the multiplexed nine different frames of the two jointly transmitted Poznan Street and Shark 3DV streams, we select the first five frames of the Poznan Street 3DV stream and the first four frames of the Shark 3DV stream to be the faces of the proposed Rubik's cube as shown in Fig. 5.7.

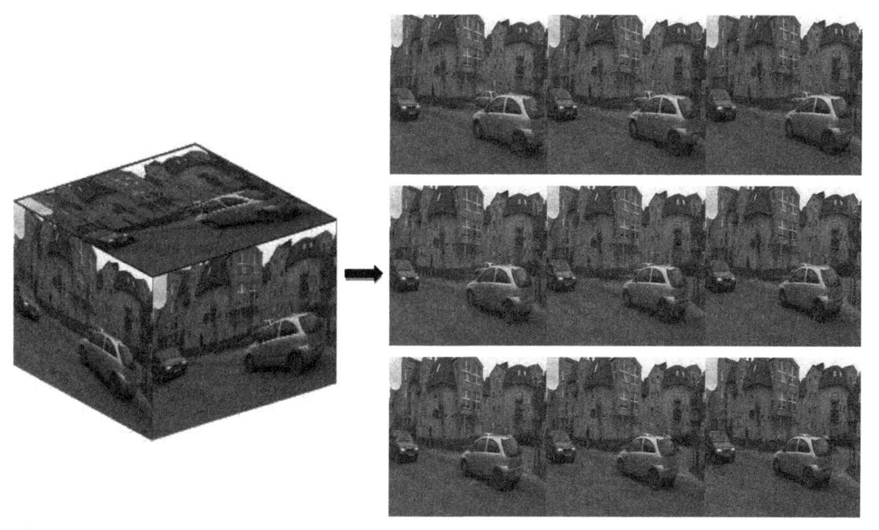

Fig. 5.5: Converted 3-D Rubik's cube faces into a 2-D frame of the tested first nine frames of the Poznan_Street 3DV stream.

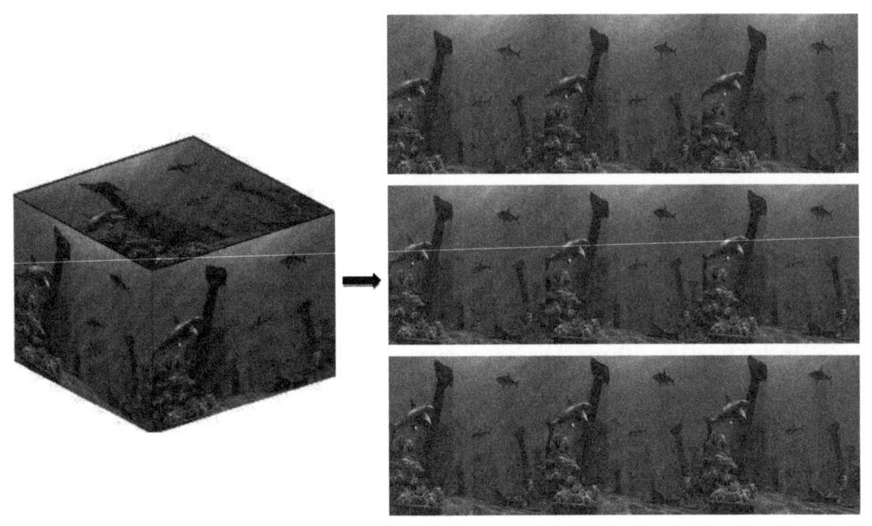

Fig. 5.6: Converted 3-D Rubik's cube faces into a 2-D frame of the tested first nine frames of the Shark 3DV stream.

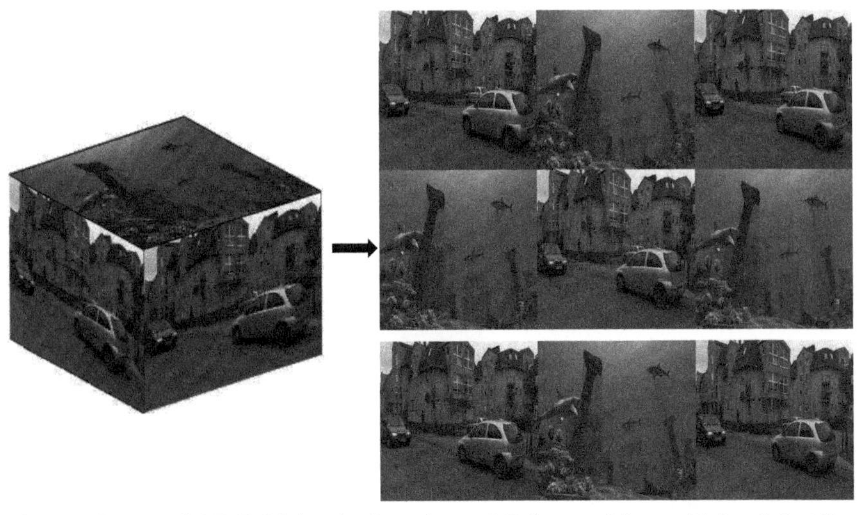

Fig. 5.7: Converted 3-D Rubik's cube faces into a 2-D frame of the multiplexed first five frames of the Poznan_Street 3DV stream and the first four frames of the Shark 3DV stream.

The third tested simulation case is the selection of nine different multiplexed frames of the nine jointly transmitted different (Ballroom, Ballet, Balloons, Dancer, Exit, Kendo, Objects2, Poznan Street, and Shark) 3DV streams, we select the first frame from each of these 3DV streams to be the faces of the proposed Rubik's cube, as shown in Fig. 5.8.

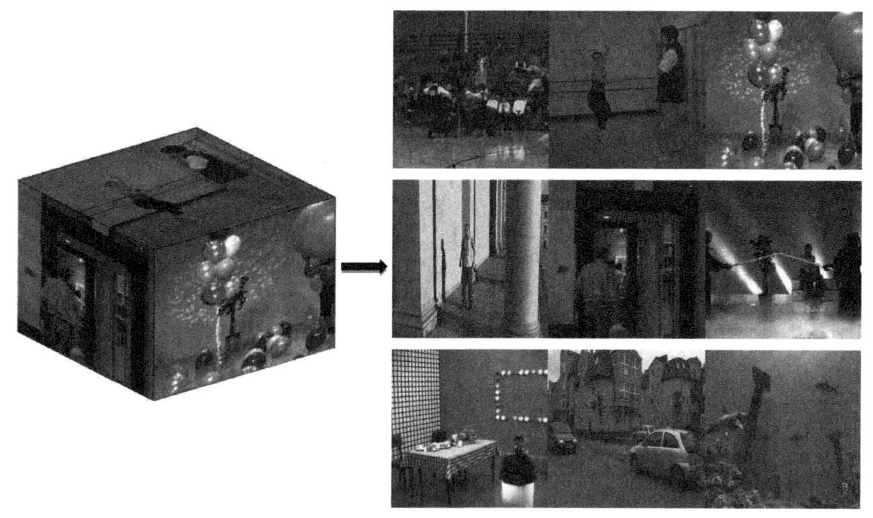

Fig. 5.8: Converted 3-D Rubik's cube faces into a 2-D frame of the multiplexed first frame in each of the Ballroom, Ballet, Balloons, Dancer, Exit, Kendo, Objects2, Poznan_Street, and Shark 3DV streams.

5.3.1 Histogram Analysis

Figures 5.9 to 5.12 show the histogram results of the three tested above-mentioned simulation cases. It is observed that the full proposed hybrid encryption framework with the Rubik's cube encryption technique gives approximately-flat histograms for all simulation tests, which means high and good quality of encryption. Therefore, the obtained results of the three tested simulation cases prove the importance of the utilization of the Rubik's cube encryption technique incorporated with the proposed hybrid encryption techniques compared to the case of not employing any encryption technique for all tested 3DV streams.

5.3.2 Deviation and Correlation Coefficient

Tables 5.1 and 5.2 show the deviation and correlation values of the nine frames of the Poznan Street and Shark 3DV sequences of the proposed hybrid encryption techniques with and without the proposed Rubik's cube encryption technique.

Table 5.3 shows the deviation and correlation values of the nine frames of the jointly encrypted Poznan Street and Shark 3DV sequences using the hybrid encryption framework with and without the proposed Rubik's cube encryption technique.

Table 5.4 shows the deviation and correlation values of the nine different frames of the jointly encrypted Ballroom, Ballet, Balloons, Dancer, Exit, Kendo, Objects2, Poznan Street, and Shark 3DV sequences of using the

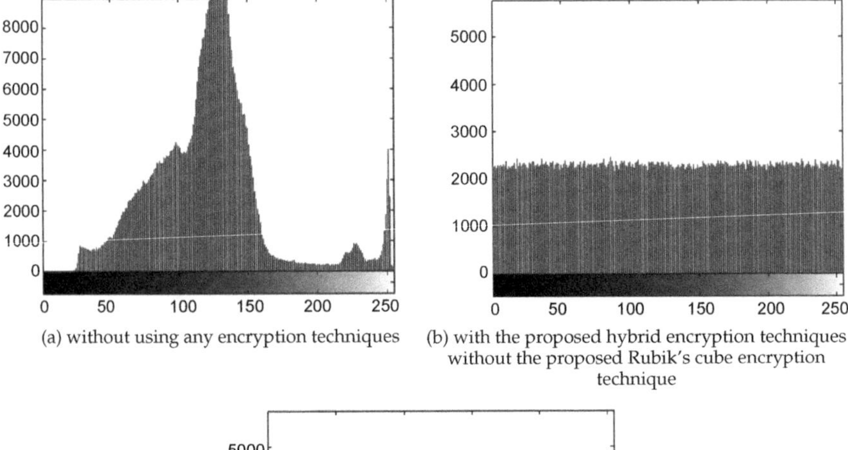

(a) without using any encryption techniques

(b) with the proposed hybrid encryption techniques without the proposed Rubik's cube encryption technique

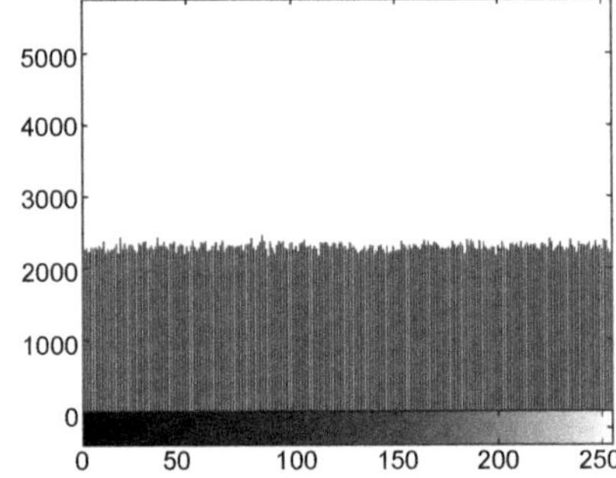

(c) with the proposed hybrid encryption techniques with the proposed Rubik's cube encryption technique

Fig. 5.9: Histograms of the encrypyed frames of the Poznan_Street 3DV stream.

hybrid encryption framework with and without the proposed Rubik's cube encryption technique.

It is noticed from the three tested simulation cases that the results of the full proposed hybrid encryption framework with the proposed Rubik's cube encryption technique are recommended and appreciated for 3DV encryption.

5.4 Simulation Results

To evaluate the performance of the suggested hybrid encryption framework, several simulation tests have been carried out on standard 3DV (Ballroom, Ballet, Balloons, Dancer, Exit, Kendo, Objects2, Poznan Street, and Shark) streams [118]. The tested 3DV sequences have different spatial and temporal characteristics. For each 3DV sequence, 8 views with 50 frames in each view are compressed. The frame rate is 20 frames per second (fps). In the simulation results, the reference JMVC codec [119] is employed depending on the H.264/AVC codec [120] to compress the 3DV sequences. The coding simulation parameters utilized in this work are selected depending on

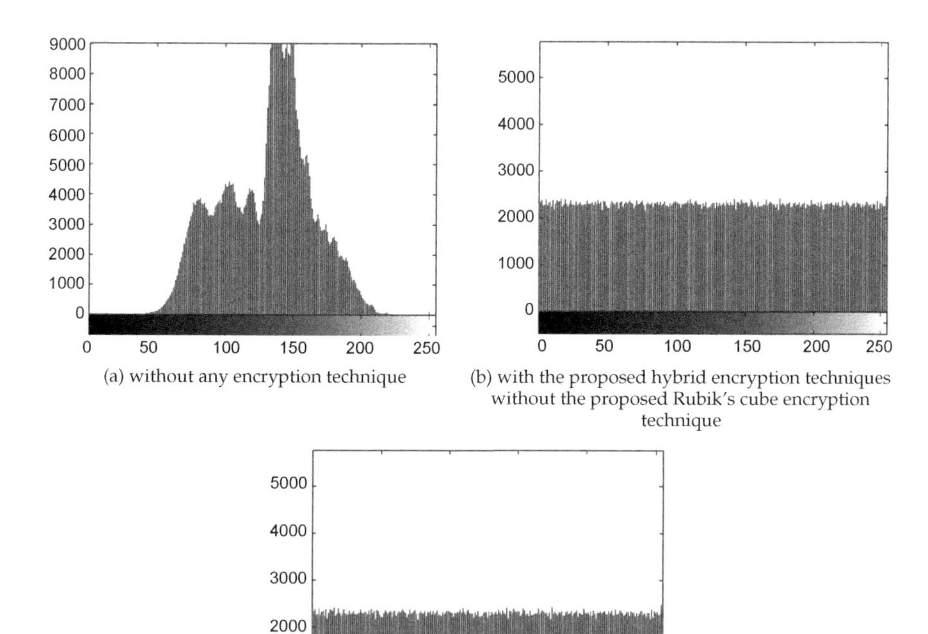

(a) without any encryption technique

(b) with the proposed hybrid encryption techniques without the proposed Rubik's cube encryption technique

(c) with the proposed hybrid encryption techniques with the proposed Rubik's cube encryption technique

Fig. 5.10: Histograms of the encrypted frames of the Shark 3DV stream.

the well-known JVT experiment [118]. For each sequence, the compressed 3-D H.264/MVC bit streams are produced, then encrypted using different AES, chaotic encryption with CFB mode, and RC6 encryption techniques with and without Rubik's cube encryption. After that, the encrypted bit streams are transported through the OFDM system. Therefore, this study is performed to simulate the AES, chaotic encryption with CFB mode, and RC6 with and without Rubik's cube method. Thus, the encrypted 3DV frames are digitized and transmitted via the OFDM communication system through AWGN and Rayleigh fading channels. In all the simulation tests, OFDM system is modulated by QPSK and 128 sub-channels, each with 500 kHz spacing. The summarized simulation parameters are like those given in Table 4.4.

To further analyze the performance of employing the suggested full hybrid encryption framework, we test the three above-mentioned simulation cases over wireless OFDM system with its three different versions of DCT-OFDM, DWT-OFDM, and FFT-OFDM. These comparison tests are implemented to prove the efficiency of the proposed hybrid encryption

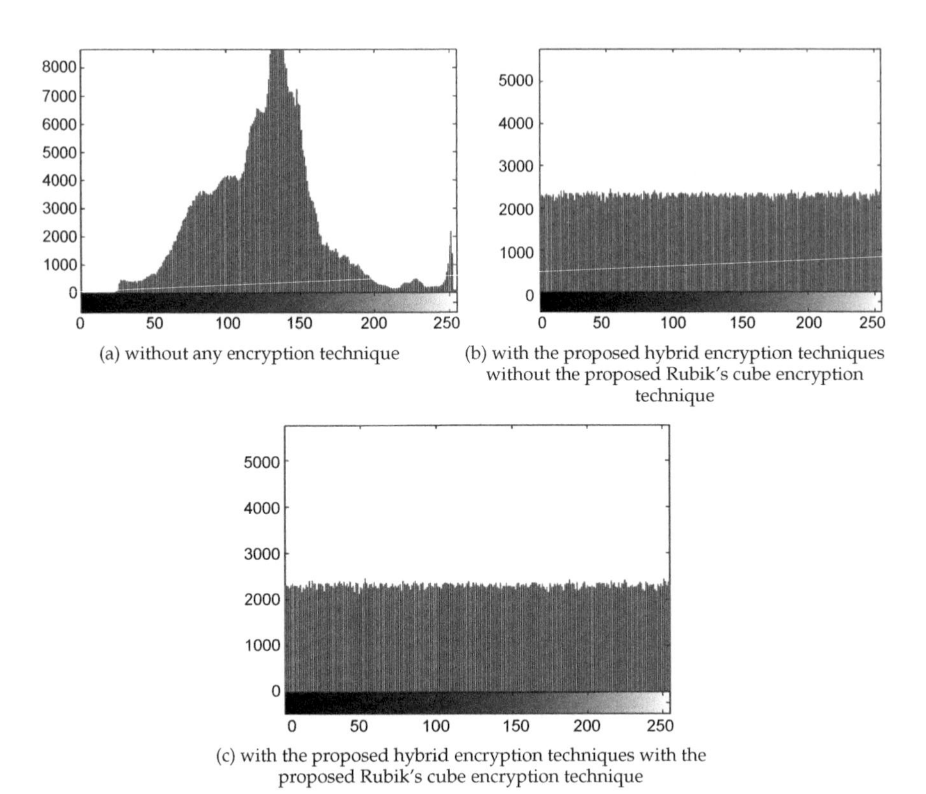

(a) without any encryption technique

(b) with the proposed hybrid encryption techniques without the proposed Rubik's cube encryption technique

(c) with the proposed hybrid encryption techniques with the proposed Rubik's cube encryption technique

Fig. 5.11: Histograms of the multiplexed encrypted frames of the Poznan_Street and Shark 3DV streams.

framework and to select the most appropriate modulation technique that can guarantee our objectives of high security of encryption and good quality of received 3DV frames. Therefore, the encrypted 3DV frames are first digitized and then transmitted via different versions of the OFDM communication system through AWGN and Rayleigh fading channels.

Table 5.5 shows the PSNR performance versus E_b/N_o for the OFDM system with its three different versions over the Rayleigh fading channel at f_d = 600 Hz in the case of the nine different frames of the Rubik's cube faces taken from the Poznan Street 3DV frames. The proposed hybrid encryption techniques are implemented with the Rubik's cube technique. Table 5.6 shows the PSNR performance vs. E_b/N_o for the OFDM system with its three different versions over the Rayleigh fading channel at f_d = 600 Hz in the case of nine different frames on the Rubik's cube faces taken from the Shark 3DV frames.

Table 5.7 shows the PSNR performance vs. E_b/N_o for the OFDM system with its three different versions over the Rayleigh fading channel at f_d = 600 Hz in the case of the nine different frames of the Rubik's cube faces taken

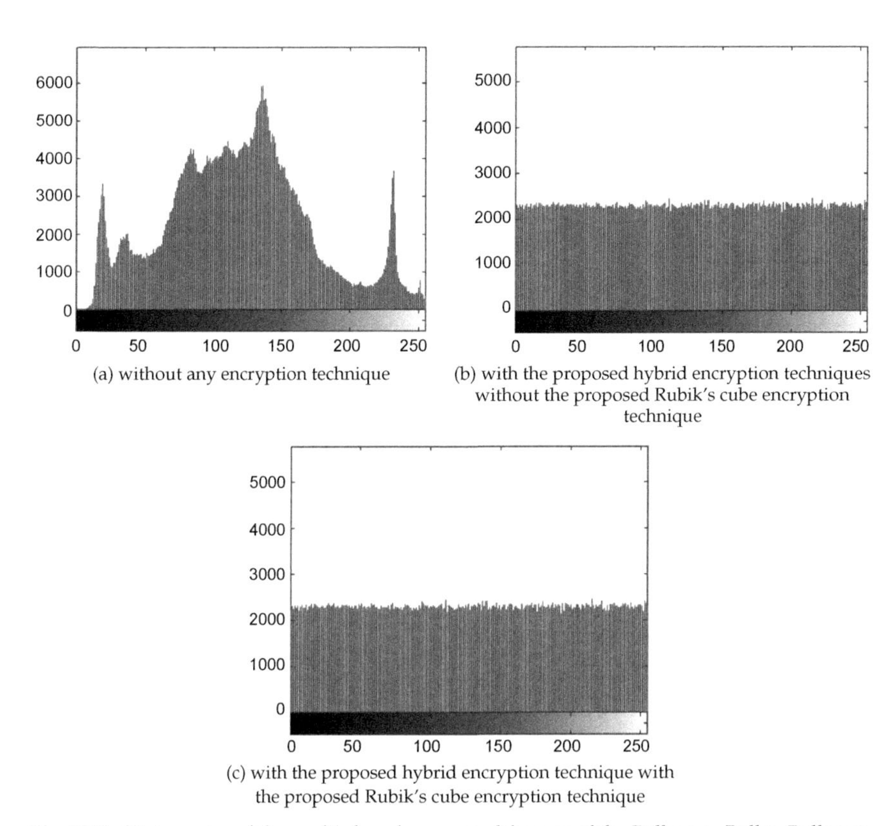

(a) without any encryption technique

(b) with the proposed hybrid encryption techniques without the proposed Rubik's cube encryption technique

(c) with the proposed hybrid encryption technique with the proposed Rubik's cube encryption technique

Fig. 5.12: Histograms of the multiplexed encrypted frames of the Ballroom, Ballet, Balloons, Dancer, Exit, Kendo, Objects2, Poznan_Street, and Shark 3DV streams.

from Poznan Street and Shark 3DV frames. The proposed hybrid encryption techniques employing the Rubik's cube encryption are considered. Table 5.8 shows the PSNR performance vs. E_b/N_o for the OFDM system in the case of the nine different frames of the Rubik's cube faces taken from Ballroom, Ballet, Balloons, Dancer, Exit, Kendo, Objects2, Poznan Street, and Shark 3DV frames. The results prove the significance of exploiting the proposed Rubik's cube encryption technique combined with the hybrid encryption techniques for achieving better performance. Also, it is noticed that the PSNR comparison results of the FFT-OFDM system are better than those of the DWT-OFDM and DCT-OFDM systems for all tested 3DV frames.

Figures 5.13 to 5.16 illustrate the subjective comparison results of the Poznan Street 3DV frame transmission through the OFDM system with its three different versions over the Rayleigh fading channel at f_d = 600 Hz at E_b/N_o = 10 dB for the proposed hybrid encryption techniques with Rubik's cube encryption. Figures 5.17 and 5.18 illustrate the subjective comparison

Table 5.1: Deviation and correlation values for the encrypted frames of the Poznan Street 3DV stream.

Frames of the Poznan Street 3DV Stream	Deviation Value (without Rubik's cube technique)	Correlation Value (without Rubik's cube technique)	Deviation (with Rubik's cube technique)	Correlation (with Rubik's cube technique)
Frame 1 encrypted with chaotic encryption with CFB mode	32.3465	–5.3674e-04	32.5557	0.0034
Frame 2 encrypted with RC6	32.3082	0.0013	32.3923	–0.0035
Frame 3 encrypted with AES	32.1944	0.0072	32.0339	0.0017
Frame 4 encrypted with RC6	32.1021	–1.8267e-04	32.1637	–9.4587
Frame 5 encrypted with chaotic encryption with CFB mode	32.1616	8.0253e-04	31.9791	4.8924e-04
Frame 6 encrypted with AES	32.4380	–3.7207e-04	32.3907	–0.0036
Frame 7 encrypted with RC6	32.0932	–0.0054	32.2784	–6.1112e-04
Frame 8 encrypted with AES	32.2423	0.0014	32.3655	–0.0041
Frame 9 encrypted with chaotic encryption with CFB mode	32.2719	–0.0014	32.2638	3.7318e-04
Average value	**32.2398**	**3.1234e-04**	**32.2692**	**–7.5495e-04**

Table 5.2: Deviation and correlation values for the encrypted frames of the Shark 3DV stream.

Frames of the Poznan_Street 3DV Stream	Deviation Value (without Rubik's cube technique)	Correlation Value (without Rubik's cube technique)	Deviation (with Rubik's cube technique)	Correlation (with Rubik's cube technique)
Frame 1 encrypted with chaotic encryption with CFB mode	37.3132	–0.0012	37.6356	0.0037
Frame 2 encrypted with RC6	36.9697	–0.0061	37.0092	–0.0048
Frame 3 encrypted with AES	36.6193	0.0049	36.6908	–0.0010
Frame 4 encrypted with RC6	35.7652	0.0054	35.9369	0.0013
Frame 5 encrypted with chaotic encryption with CFB mode	35.7639	0.0020	35.5714	0.0020
Frame 6 encrypted with AES	35.3974	0.0028	35.3171	0.0038
Frame 7 encrypted with RC6	35.0501	–1.5915e-04	35.1172	–5.2711e-04
Frame 8 encrypted with AES	34.8716	3.4871e-04	34.7159	0.0020
Frame 9 encrypted with chaotic encryption with CFB mode	34.5526	–0.0027	34.3853	–0.0054
Average value	**35.8114**	**5.8773e-04**	**35.8199**	**1.1921e-04**

Table 5.3: Deviation and correlation values for the multiplexed encrypted frames of the Poznan Street and Shark 3DV streams.

Frames of the Poznan Street 3DV Stream	Deviation Value (without Rubik's cube technique)	Correlation Value (without Rubik's cube technique)	Deviation (with Rubik's cube technique)	Correlation (with Rubik's cube technique)
Frame 1 encrypted with chaotic encryption with CFB mode	32.3465	–5.3674e-04	32.5835	–0.0018
Frame 2 encrypted with RC6	36.9697	–0.0061	36.8355	–0.0013
Frame 3 encrypted with AES	32.1944	0.0072	32.1450	0.0027
Frame 4 encrypted with RC6	35.7652	0.0054	35.8473	0.0033
Frame 5 encrypted with chaotic encryption with CFB mode	32.1616	8.0253e-04	32.0616	–0.0022
Frame 6 encrypted with AES	35.3974	0.0028	35.5415	0.0046
Frame 7 encrypted with RC6	32.0932	–0.0054	32.1125	0.0012
Frame 8 encrypted with AES	34.8716	3.4871e-04	34.6374	0.0054
Frame 9 encrypted with chaotic encryption with CFB mode	32.2719	–0.0014	32.3123	–0.0067
Average value	**33.7857**	**3.4606e-04**	**33.7863**	**5.7778e-04**

Table 5.4: Deviation and correlation values for the multiplexed encrypted frames of the Ballroom, Ballet, Balloons, Dancer, Exit, Kendo, Objects2, Poznan Street, and Shark 3DV streams.

Frames of the Poznan Street 3DV Stream	Deviation Value (without Rubik's cube technique)	Correlation Value (without Rubik's cube technique)	Deviation (with Rubik's cube technique)	Correlation (with Rubik's cube technique)
Frame 1 encrypted with chaotic encryption with CFB mode	20.6659	–9.7732e-04	20.7495	–7.8128e-04
Frame 2 encrypted with RC6	24.5501	–0.0031	24.6124	–0.0025
Frame 3 encrypted with AES	43.2511	0.0064	43.4819	–2.6851e-04
Frame 4 encrypted with RC6	52.0962	0.0046	52.2789	–0.0030
Frame 5 encrypted with chaotic encryption with CFB mode	12.9203	–0.0037	12.6637	–0.0025
Frame 6 encrypted with AES	36.0365	–0.0027	35.6840	0.0047
Frame 7 encrypted with RC6	41.6556	0.0057	41.9126	–0.0025
Frame 8 encrypted with AES	32.3546	–0.0054	32.3279	–0.0055
Frame 9 encrypted with chaotic encryption with CFB mode	37.0304	–0.0016	37.3102	–0.0059
Average value	**33.3956**	**–8.6369e-05**	**33.4468**	**–0.0020**

Table 5.5: PSNR performance vs. E_b/N_o for the jointly encrypted frames with the proposed hybrid encryption techniques employing Rubik's cube encryption of the Poznan Street 3DV stream, and then transmission via FFT/DCT/DWT system over Rayleigh fading channel at $f_d = 600$ Hz.

Frames of the Poznan Street 3DV Stream	FFT-OFDM (E_b/N_o = 2 dB)	FFT-OFDM (E_b/N_o = 6 dB)	FFT-OFDM (E_b/N_o = 10 dB)	DCT-OFDM (E_b/N_o = 2 dB)	DCT-OFDM (E_b/N_o = 6 dB)	DCT-OFDM (E_b/N_o = 10 dB)	DWT-OFDM (E_b/N_o = 2 dB)	DWT-OFDM (E_b/N_o = 6 dB)	DWT-OFDM (E_b/N_o = 10 dB)
Frame 1 encrypted with chaotic encryption with CFB mode	37.75	Inf.	Inf.	30.68	39.71	Inf.	30.240	41.17	Inf.
Frame 2 encrypted with RC6	31.53	45.1545	Inf.	29.35	34.94	37.75	27.91	34.01	45.15
Frame 3 encrypted with AES	31.63	Inf.	Inf.	28.67	34.36	42.14	28.16	33.39	45.15
Frame 4 encrypted with RC6	32.03	Inf.	Inf.	29.47	34.74	38.62	28.52	33.11	48.16
Frame 5 encrypted with chaotic encryption with CFB mode	39.71	Inf.	Inf.	30.03	37.75	48.16	29.77	40.38	Inf.
Frame 6 encrypted with AES	31.63	Inf.	Inf.	28.87	34.01	41.17	28.76	33.25	Inf.
Frame 7 encrypted with RC6	32.48	Inf.	Inf.	29.71	34.01	38.16	27.95	33.39	45.15
Frame 8 encrypted with AES	32.60	Inf.	Inf.	29.18	33.85	38.62	27.59	33.85	48.16
Frame 9 encrypted with chaotic encryption with CFB mode	35.15	Inf.	Inf.	31.26	38.16	43.39	30.38	40.38	Inf.

Table 5.6: PSNR performance vs. E_b/N_o for the jointly encrypted frames with the proposed hybrid encryption techniques employing Rubik's cube encryption of the Shark 3DV stream, and then transmission via FFT/DCT/DWT system over Rayleigh fading channel at $f_d = 600$ Hz.

Frames of the Shark 3DV Stream	FFT-OFDM ($E_b/N_o = 2$ dB)	FFT-OFDM ($E_b/N_o = 6$ dB)	FFT-OFDM ($E_b/N_o = 10$ dB)	DCT-OFDM ($E_b/N_o = 2$ dB)	DCT-OFDM ($E_b/N_o = 6$ dB)	DCT-OFDM ($E_b/N_o = 10$dB)	DWT-OFDM ($E_b/N_o = 2$ dB)	DWT-OFDM ($E_b/N_o = 6$ dB)	DWT-OFDM ($E_b/N_o = 10$ dB)
Frame 1 encrypted with chaotic encryption with CFB mode	39.71	Inf.	Inf.	32.60	43.39	Inf.	32.36	48.16	Inf.
Frame 2 encrypted with RC6	29.83	48.1648	Inf.	28.72	34.01	37.02	26.73	31.44	Inf.
Frame 3 encrypted with AES	29.65	Inf.	Inf.	28.38	32.03	37.37	26.49	31.73	Inf.
Frame 4 encrypted with RC6	29.96	48.1648	Inf.	28.20	32.25	39.13	27.12	32.60	45.15
Frame 5 encrypted with chaotic encryption with CFB mode	39.71	Inf.	Inf.	32.97	42.14	Inf.	31.44	48.16	Inf.
Frame 6 encrypted with AES	29.90	48.16	Inf.	28.57	32.14	35.15	27.05	31.73	48.16
Frame 7 encrypted with RC6	31.26	Inf.	Inf.	28.34	33.69	39.71	26.61	31.83	42.14
Frame 8 encrypted with AES	30.68	Inf.	Inf.	28.29	33.25	37.75	26.79	32.72	39.71
Frame 9 encrypted with chaotic encryption with CFB mode	40.38	Inf.	Inf.	31.63	42.14	48.16	32.48	48.16	Inf.

Table 5.7: PSNR performance vs. E_b/N_o for the jointly encrypted frames with the proposed hybrid encryption techniques using Rubik's cube encryption of the Poznan Street and Shark 3DV streams, and then transmission via FFT/DCT/DWT system over Rayleigh fading channel at f_d = 600 Hz.

Multiplexed Frames of the Poznan Street and Shark 3DV Streams	**FFT-OFDM (E_b/N_o = 2 dB)**	**FFT-OFDM (E_b/N_o = 6 dB)**	**FFT-OFDM (E_b/N_o = 10 dB)**	**DCT-OFDM (E_b/N_o = 2 dB)**	**DCT-OFDM (E_b/N_o = 6 dB)**	**DCT-OFDM (E_b/N_o = 10 dB)**	**DWT-OFDM (E_b/N_o = 2 dB)**	**DWT-OFDM (E_b/N_o = 6 dB)**	**DWT-OFDM (E_b/N_o = 10 dB)**
Frame 1 encrypted with chaotic encryption with CFB mode	37.75	Inf.	Inf.	30.68	39.71	Inf.	30.240	41.17	Inf.
Frame 2 encrypted with RC6	29.83	48.1648	Inf.	28.72	34.01	37.02	26.73	31.44	Inf.
Frame 3 encrypted with AES	31.63	Inf.	Inf.	28.67	34.36	42.14	28.16	33.39	45.15
Frame 4 encrypted with RC6	29.96	48.1648	Inf.	28.20	32.25	39.13	27.12	32.60	45.15
Frame 5 encrypted with chaotic encryption with CFB mode	39.71	Inf.	Inf.	30.03	37.75	48.16	29.77	40.38	Inf.
Frame 6 encrypted with AES	29.90	48.16	Inf.	32.97	42.14	Inf.	31.44	48.16	Inf.
Frame 7 encrypted with RC6	32.48	Inf.	Inf.	29.71	34.01	38.16	27.95	33.39	45.15
Frame 8 encrypted with AES	30.68	Inf.	Inf.	28.34	33.69	39.71	26.61	31.83	42.14
Frame 9 encrypted with chaotic encryption with CFB mode	35.15	Inf.	Inf.	31.26	38.16	43.39	30.38	40.38	Inf.

Table 5.8: PSNR performance vs. E_b/N_o for the jointly encrypted frames with the proposed hybrid encryption techniques using Rubik's cube encryption of the Ballroom, Ballet, Balloons, Dancer, Exit, Kendo, Objects2, Poznan Street, and Shark streams, and then transmission via FFT/DCT/DWT system over Rayleigh fading channel at $f_d = 600$ Hz.

Multiplexed Frames of the Nine Different 3DV Streams	FFT-OFDM (E_b/N_o = 2 dB)	FFT-OFDM (E_b/N_o = 6 dB)	FFT-OFDM (E_b/N_o = 10 dB)	DCT-OFDM (E_b/N_o = 2 dB)	DCT-OFDM (E_b/N_o = 6 dB)	DCT-OFDM (E_b/N_o = 10 dB)	DWT-OFDM (E_b/N_o = 2 dB)	DWT-OFDM (E_b/N_o = 6 dB)	DWT-OFDM (E_b/N_o = 10 dB)
Frame 1 encrypted with chaotic encryption with CFB mode	45.15	Inf.	Inf.	36.70	48.16	Inf.	36.70	48.16	Inf.
Frame 2 encrypted with RC6	33.54	Inf.	Inf.	30.84	36.40	39.13	31.17	35.37	48.16
Frame 3 encrypted with AES	30.45	48.16	Inf.	27.19	31.08	36.12	25.96	31.83	42.14
Frame 4 encrypted with RC6	30.03	Inf.	Inf.	28.38	32.14	36.12	26.64	32.48	48.16
Frame 5 encrypted with chaotic encryption with CFB mode	43.39	Inf.	Inf.	37.02	48.16	Inf.	34.74	48.16	Inf.
Frame 6 encrypted with AES	32.14	Inf.	Inf.	29.96	35.61	38.62	28.76	33.85	48.16
Frame 7 encrypted with RC6	30.68	Inf.	Inf.	28.03	32.03	37.37	26.79	32.60	41.17
Frame 8 encrypted with AES	31.83	Inf.	Inf.	28.87	34.94	39.13	28.81	33.25	Inf.
Frame 9 encrypted with chaotic encryption with CFB mode	36.40	Inf.	Inf.	31.35	45.15	48.16	34.36	45.15	Inf.

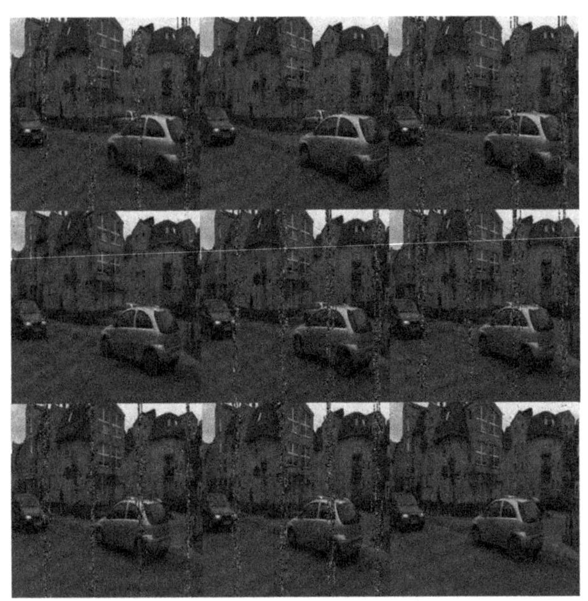

Fig. 5.13: Subjective comparison results of the Poznan Street 3DV frames transmitted through the DCT-OFDM system over the Rayleigh fading channel at f_d = 600 Hz at E_b/N_o = 10 dB with the proposed hybrid encryption techniques employing the Rubik's cube encryption.

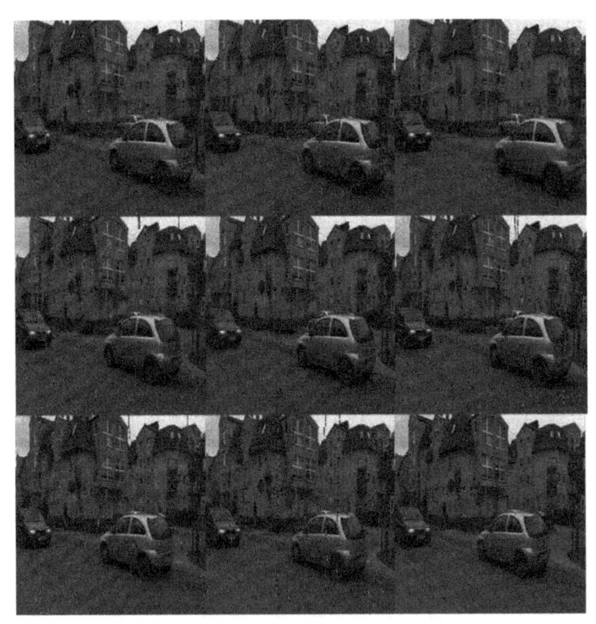

Fig. 5.14: Subjective comparison results of the Poznan Street 3DV frames transmitted through the DWT-OFDM system over the Rayleigh fading channel at f_d = 600 Hz at E_b/N_o = 10 dB with the proposed hybrid encryption techniques employing the Rubik's cube encryption.

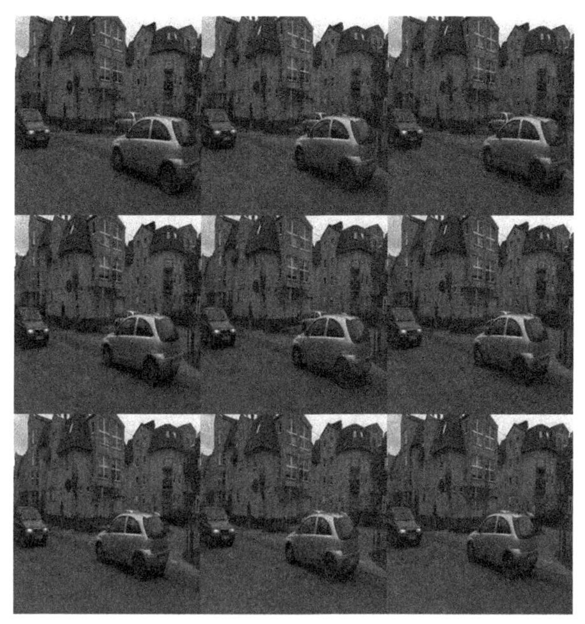

Fig. 5.15: Subjective comparison results of the Poznan Street 3DV frames transmitted through the FFT-OFDM system over the Rayleigh fading channel at f_d = 600 Hz at E_b/N_o = 10 dB with the proposed hybrid encryption techniques employing the Rubik's cube encryption.

Fig. 5.16: Subjective comparison results of the jointly encrypted Ballroom, Ballet, Balloons, Dancer, Exit, Kendo, Objects2, Poznan Street, and Shark 3DV frames with transmission through the DCT-OFDM system over the Rayleigh fading channel at f_d = 600 Hz and E_b/N_o = 10 dB.

Fig. 5.17: Subjective comparison results of the jointly encrypted Ballroom, Ballet, Balloons, Dancer, Exit, Kendo, Objects2, Poznan Street, and Shark 3DV frames with transmission through the DWT-OFDM system over the Rayleigh fading channel at $f_d = 600$ Hz and $E_b/N_o = 10$ dB.

Fig. 5.18: Subjective comparison results of the jointly encrypted Ballroom, Ballet, Balloons, Dancer, Exit, Kendo, Objects2, Poznan Street, and Shark 3DV frames with transmission through the FFT-OFDM system over the Rayleigh fading channel at $f_d = 600$ Hz and $E_b/N_o = 10$ dB.

results of the jointly encrypted Ballroom, Ballet, Balloons, Dancer, Exit, Kendo, Objects2, Poznan Street, and Shark 3DV frames. Transmission is performed through the OFDM system with its three different versions over the Rayleigh fading channel at $f_d = 600$ Hz at $E_b/N_o = 10$ dB considering the proposed hybrid encryption techniques with the Rubik's cube encryption. It is observed from the subjective comparison results that the performance of the FFT-OFDM system is better than those of the DCT-OFDM and DWT-OFDM systems for all tested fames, where the decrypted frames are received typically similar to the original ones with high quality.

From all presented objective and subjective results, it is recognized that the proposed hybrid encryption techniques with the Rubik's cube encryption are more appreciated and recommended for 3DV transmission. The full proposed hybrid encryption framework has preferable objective and visual simulation results compared to the case of not exploiting the proposed Rubik's cube technique. Furthermore, it is noticed that the suggested hybrid encryption framework offers adequate experimental results for various standard 3DV videos with different spatial and temporal features.

Chapter 6

Conclusions and Future Work

6.1 Conclusion

This book was devoted to the study of encrypted image and video transmission over wireless systems, like satellite and mobile systems. The book presented a study of diffusion and permutation-based image encryption. A comparison study revealed that diffusion-based encryption is more secure than permutation-based encryption from the cryptographic point of view. From the noise immunity point of view, permutation-based algorithms are more robust.

The book has also presented an enhanced chaotic encryption scheme based on different modes of operation to enhance the security of chaotic encryption. This enhancement in security has been revealed with uniform histograms. Another enhancement in diffusion-based algorithms has also been presented in the book. This enhancement is based on pre-processing of the data to allow encryption in the ECB mode, which is the simplest encryption mode, but it has the parallel processing advantage. With the proposed pre-processing network, encryption in the ECB mode with a large degree of security and parallel processing is possible. This advantage can be easily used for encrypting images and videos with few details.

The different versions of OFDM have been studied for the transmission of encrypted images and videos with all encryption schemes. Simulation results revealed that FFT-OFDM has achieved the best performance among all OFDM versions for the transmission of encrypted images and videos. Results have also revealed that chaotic encryption is better than diffusion-based encryption for transmission over wireless channels due to the immunity of chaotic encryption to noise.

Simulation experiments have revealed that the proposed enhanced encryption algorithms can perform well over wireless channels. These algorithms have achieved a good encryption quality in addition to a good performance in the transmission over wireless channels.

The effects of channel equalization, CFO compensation, and PAPR reduction methods have been studied in this book. Simulation results have shown that PSE and equalization have enhanced the performance of all image transmission schemes. In addition, results have shown that CFO compensation and PAPR reduction are necessary for good performance in all images and video transmission systems.

6.2 Future Work

The future work in this book may include the following points:

1. Investigation of different types of chaotic maps for data encryption; either 1-D, 2-D, or 3-D.
2. Implementation of image encryption in transform domains to make use of the sophisticated characteristics of all transforms.
3. Transmission of watermarked images with encrypted watermarks.
4. Study of the effect of phase noise on the image transmission systems.
5. Study of continuous phase modulation-based OFDM for possible use of the transmission of encrypted images.
6. Study of the peak-to-average power ratio reduction problem and its effect on image transmission.

References

1. S.J. Lee, J.S. Yoon and H.K. Song. May, 2009. Frequency offset mitigation of cooperative OFDM system in wireless digital broadcasting. IEEE Transactions on Consumer Electronics 55(2).
2. S. Alvarez, J. Chen, D. Lecumberri and C.P. Yang. October, 1999. HDTV: The Engineering History.
3. Mai Helmy, El-Sayed M. El-Rabaie, Ibrahim M. Eldokany and Fathi E. Abd El-Samie. November, 2017. 3-D Image Encryption Based on Rubik's Cube and RC6 Algorithm. Springer-3-D Research.
4. Mai Helmy, W. El-Shafai, S. El-Rabaie, I.M. El-Dokany, M. El-Halawany and Fathi E. Abd El-Samie. October, 2019. Proposed hybrid encryption framework for robust 3-D video wireless communication. Multimedia Tools and Applications, Springer.
5. R.V. Nee and R. Prasad. 2000. OFDM for Wireless Multimedia Communications. Artech House.
6. H. Dai and H.V. Poor. May, 2003. Advanced signal processing for power line communications. IEEE Commun. Mag. 41(5): 100–107.
7. S. Moriyama, K. Tsuchida and M. Sasaki. March, 1998. Digital transmission of high bit rate signals using 16DAPSK-OFDM modulation scheme. IEEE Transactions on Broadcasting 44(I).
8. M. Yang, N. Bourbakis and S. Li. August–September, 2004. Data-image-video encryption. IEEE, IEEE Potentials 23(3).
9. Fathi E. Abd El-Samie, Hossam Eldin H. Ahmed, Ibrahim F. Elashry, Mai H. Shahieen, Osama S. Faragallah, El-Sayed M. El-Rabaie and Saleh A. Alshebeili. October 23, 2015. Image Encryption: A Communication Perspective. Taylor & Francis eBooks, 418 pp.
10. J. Daemen and V. Rijmen. March, 1999. The Rijndael Block Cipher. AES Proposal: Rijndael, Document version 2.
11. J.J. Buchholz. December, 2001. Matlab Implementation of the Advanced Encryption Standard. http://buchholz.hs-bremen.de.
12. N. El-Fishawy and O.M. Abu Zaid. November, 2007. Quality of encryption measurement of bitmap images with RC6, MRC6, and Rijndael block cipher algorithms. International Journal of Network Security 5(3): 241–251.
13. D. Coppersmith. May, 1994. The Data Encryption Standard (DES) and its strength against attacks. IBM J. Res. Develop. 38(3).
14. G.H. Kim, J.N. Kim and G.Y. Cho. February, 2009. An improved RC6 algorithm with the same structure of encryption and decryption. ICACT.
15. M.S. Liu, Y. Zhang and J. Huali. July, 2009. Research on improving security of DES by chaotic mapping. Proceedings of the Eighth International Conference on Machine Learning and Cybernetics, Baoding, 12–15.
16. J. Daemen and V. Rijmen. 1998. The Block Cipher Rijndael, Smart Card Research and Applications. LNCS 1820, Springer-Verlag, pp. 288–296.
17. Federal Information Processing Standards Publication 197, Advanced Encryption Standard, November, 2001.

18. J. Fridrich. 1998. Symmetric ciphers based on two-dimensional chaotic maps. Int. J. Bifurcation Chaos Appl. Sci. Eng., 1259–1284.
19. F. Dachselt and W. Schwarz. December, 2001. Chaos and cryptography. IEEE Transactions on Circuits and Systems 48(12).
20. M. Asiml and V. Jeotil. February 22–24, 2007. On Image Encryption: Comparison Between AES and a Novel Chaotic Encryption Scheme. IEEE-ICSCN 2007, MIT Campus, Anna University, Chennai, India, pp. 65–69.
21. C.S. Avilaf and R.S. Reillot. 2001. The Rijndael Block Cipher (AES Proposal): A Comparison with DES. Proceedings IEEE 35th Annual 2001 International Carnahan Conference on Security Technology (Cat. No.01CH37186).
22. R.L. Rivest, M.J.B. Robshaw, R. Sidney and Y.L. Yin. 1998. The RC6™ Block Cipher. M.I.T. Laboratory for Computer Science, 545 Technology Square, Cambridge, MA 02139, USA.
23. Y.C. Chen and L.W. Chang. 2001. A secure and robust digital watermarking technique by the block cipher RC6 and secure hash algorithm. Proceedings 2001 International Conference on Image Processing (Cat. No.01CH37205).
24. S. Liu, J. Sun, Z. Xu and J. Liu. 2008. Analysis on an image encryption algorithm. IEEE International Workshop on Education Technology and Training & 2008 International Workshop on Geoscience and Remote Sensing.
25. L. Kocarev. 2001. Chaos-based cryptography: A brief overview. IEEE Circ. Syst. Mag. 1(3): 6–21.
26. D. Kim and G.L. Stüber. October, 1998. Residual ISI cancellation for OFDM with applications to HDTV broadcasting. IEEE Journal on Selected Areas in Communications 16(8).
27. G.H. Yang, D. Shen and V.O.K. Li. 2004. UEP for video transmission in space-time coded OFDM systems. IEEE INFOCOM.
28. M.I. Rahman, S.S. Das and Frank H.P. Fitzek. February, 2005. OFDM based WLAN systems. Center for TeleInFrastruktur (CTiF), Technical Report R-04-1002; v1.2, 18.
29. E.P. Lawrey. December, 2001. Adaptive Techniques for Multiuser OFDM. PHd Thesis, James Cook University.
30. E. Lawrey. October, 1997. The Suitability of OFDM as a Modulation Technique for Wireless Telecommunications, with a CDMA Comparison, Bachelor Thesis. James Cook University.
31. K. Abdullah and Z.M. Hussain. December 2–5, 2007. Performance of fourier-based and wavelet-based OFDM for DVB-T systems. In the proceeding of the 2007 Australasian Telecommunication Networks and Applications Conference, Christchurch, New Zealand.
32. J.J. van de Beek, O. Edfors, M. Sandell, S.K. Wilson and P.O. Borjesson. 1995. On channel estimation in OFDM systems. Vehicular Technology Conference, IEEE 45th.
33. P. Tan and N.C. Beaulieu. November, 2006. A comparison of DCT-based OFDM and DFT-based OFDM in frequency offset and fading channels. IEEE Trans. Communs. 54(11).
34. J.G. Andrews, A. Ghosh and R. Muhamed. February, 2007. Fundamentals of WiMAX understanding broadband wireless networking. Prentice Hall Communications Engineering and Emerging Technologies Series, pp. 113–145.
35. A.K. Lee, Ooi, M. Drieberg and V. Jeoti. 2006. DWT based FFT in practical OFDM systems. TENCON 2006–2006 IEEE Region 10 Conference.
36. P. Liu, B.B. Li, Z.Y. Lu and F.K. Gong. 2005. An OFDM Bandwidth Estimation Scheme for Spectrum Monitoring. Proceedings 2005 International Conference on Wireless Communications, Networking and Mobile Computing.
37. X.F. Wang, Y.R. Shayan and M. Zeng. November, 2004. On the code and interleaver design of broadband OFDM systems. IEEE Communications Letters 8(11).
38. F. Gao, T. Cui, A. Nallanathan and C. Tellambura. September, 2008. Maximum likelihood based estimation of frequency and phase offset in DCT OFDM systems under non-circular transmissions: Algorithms, analysis and comparisons. IEEE Trans. Communs. 56(9).
39. R. Merched. August, 2006. On OFDM and signal-carrier frequency-domain systems based on trigonometric transforms. IEEE Signal Process. Letter 13(8).

40. E. Lawrey, C.J. Kikkert. August, 1999. Peak to average power ratio reduction of OFDM signals using peak reduction carriers. Fifth International Symposium on Signal Processing and its Applications, ISSPA '99, Brisbane, Australia, 22–25.
41. M. Shen, G. Li and H. Liu. July, 2005. Effect of traffic channel configuration on the orthogonal frequency division multiple access downlink performance. IEEE Transactions on Wireless Communications 4(4).
42. H. Schulze and C. Luders. 2005. Theory and Application of OFDM and CDMA Wideband Wireless Communications. John Wiley & Sons Ltd., pp. 145–264.
43. N. Al-Dhahir and H. Minn. March 13–17, 2005. A new multicarrier transceiver based on the discrete cosine transform. In Proc. of the IEEE Wireless Commun. and Net. Conf. 1: 45–50.
44. P. Tan and N.C. Beaulieu. 2005. An improved DCT-based OFDM data transmission scheme. In the proceeding of the IEEE 16th PIMRC'05.
45. H. Harada and R. Prasad. 2002. Simulation and software radio for mobile communications. House Universal Personal Communications Library.
46. M.S. El-Tanany, Y. Wu and L. Házy. March, 2001. OFDM uplink for interactive broadband wireless: Analysis and simulation in the presence of carrier, clock and timing errors. IEEE Transactions on Broadcasting 47(1).
47. A. Langowski. 2009. Time and frequency synchronisation in 4-G OFDM systems. Hindawi Publishing Corporation EURASIP Journal on Wireless Communications and Networking Volume.
48. Mary Ann Ingram. August, 2000. OFDM Simulation Using Matlab. Smart Antenna Research Laboratory, Guillermo Acosta.
49. J. Zhang and B. Li. 2008. New modulation identification scheme for OFDM in multipath Rayleigh fading channel. International Symposium on Computer Science and Computational Technology.
50. H.S. Chu, B.S. Park, C.K. An, J.S. Kang and H.G. Son. 2007. Wireless Image Transmission based on Adaptive OFDM System. 2007 International Forum on Strategic Technology.
51. P. Tan and N.C. Beaulieu. 2005. Precise bit error probability analysis of DCT OFDM in the presence of carrier frequency offset on AWGN channels. In the proceeding of the IEEE Globcom, pp. 1429–1434.
52. Giridhar D. Mandyam. 2003. On the discrete cosine transform and OFDM systems. Nokia Research Center. 6000 Connection Drive, Irving, TX 75039 USA, IEEE, ICASSP.
53. V.B. Vats, K.K. Garg and A. Abad. 2008. Performance analysis of DFT-OFDM, DCT-OFDM, and DWT-OFDM systems in AWGN. In the proceeding of the IEEE Fourth International Conference on Wireless and Mobile Communications.
54. M. Misiti, Y. Misiti and G. Oppenheim. 2007. Wavelets and their applications. Jean-Michel Poggi published in Great Britain and the United States by ISTE Ltd.
55. A. Prochazka, J. Uhlir, P.J.W. Rayner and N.J. Kingsbury. 1998. Signal Analysis and Prediction. Birkhauser Inc.
56. B. Muquet, Z. Wang, G.B. Giannakis, M.d. Courville and P. Duhamel. December, 2002. Cyclic prefixing or zero padding for wireless multicarrier transmissions. IEEE Transactions on Communications 50(12).
57. B. Li, S. Zhou, M. Stojanovic, L. Freitag and P. Willett. 2007. Non-uniform doppler compensation for zero-padded OFDM over fast-varying under water acoustic channels. OCEANS 2007–Europe, IEEE.
58. C.R.N. Athaudage and R.R.V. Angiras. December, 2005. Sensitivity of FFT-equalised zero-padded OFDM systems to time and frequency synchronization errors. IEE Proc. Commun. 152(6).
59. B. Muquet, M.D. Courville, P. Duhamel, G.B. Giannakis and P. Magniez. November, 2002. Turbo demodulation of zero-padded OFDM transmissions. IEEE Transactions on Communications 50(11).
60. D. Huang and K.B. Letaief. July, 2005. An interference-cancellation scheme for carrier frequency offsets correction in OFDMA systems. IEEE Trans. Commun. 53(7): 1155–1165.

61. J. Paul and M.G. Linnartz. November, 2001. Performance analysis of synchronous MC-CDMA in mobile rayleigh channel with both delay and doppler spreads. IEEE Transactions on Vehicular Technology 50(6).
62. P. Tan and N.C. Beaulieu. 2005. Precise bit error probability analysis of DCT OFDM in the presence of carrier frequency offset on AWGN channels. In Proceedings of the IEEE Globcom, pp. 1429–1434.
63. Lawrey, E.P. 2001. Adaptive techniques for multiuser OFDM. Ph.D. Thesis, James Cook University.
64. Ashwaq T. Hashim, Janan A. Mahdi and Salma H. Abdullah. 2010. A proposed 512 bits RC6 encryption algorithm. IJCCCE 10(1).
65. S. Li, X. Mou and Y. Cai. 2001. Improving security of a chaotic encryption approach. Physics Letters A 290(3-4): 127–133.
66. Fouad Ramia and G.M.U. Hunar Qadir. December, 2006. RC6 implementation including key scheduling using FPGA. ECE 646 Project.
67. A.B. Mohamed, G. Zaibi and A. Kachouri. March, 2011. Implementation of RC5 and RC6 block ciphers on digital images. Eighth International Multi-Conference on Systems, Signals & Devices.
68. Harsh Kumar Verma and Ravindra Kumar Singh. March, 2012. Performance analysis of RC6, twofish and Rijndael block cipher algorithms. International Journal of Computer Applications (0975–8887) 42(16).
69. Mansoor Ebrahim, Karachi, Shujaat Khan, Karachi and Umer Bin Khalid. January, 2013. Symmetric algorithm survey: a comparative analysis. International Journal of Computer Applications (0975–8887) 61(20).
70. M. Sirisha and S.V.V.S. Lakshmi. May, 2014. Pixel transformation based on Rubik's cube principle. International Journal of Application or Innovation in Engineering & Management (IJAIEM) 3(5).
71. S. Li, X. Zheng, X. Mou and Y. Cai. 2002. Chaotic encryption scheme for real-time digital video. In Proceedings of SPIE 4666: 149–160.
72. ISO/IEC JTC1/SC29/WG11. 2006. Common test conditions for multiview video coding. JVT-U207, Hangzhou, China.
73. http://wftp3.itu.int/av-arch/jvt-site/2009_01_Geneva/JVT. zip: Reference software for multiview video coding (mvc). Last accessed on 25/10/2016.
74. Lini Abraham and Neenu Daniel. April, 2013. Secure image encryption algorithms: A review. International Journal of Scientific & Technology Research 2(4).
75. Mansoor Ebrahim, Karachi, Shujaat Khan, Karachi and Umer Bin Khalid. January, 2013. Symmetric algorithm survey: A comparative analysis. International Journal of Computer Applications 61(20): 0975–8887.
76. M. Sirisha and S.V.V.S. Lakshmi. May, 2014. Pixel transformation based on Rubik's cube principle. International Journal of Application or Innovation in Engineering & Management (IJAIEM) 3(5).
77. S. Li, X. Zheng, X. Mou and Y. Cai. 2002. Chaotic encryption scheme for real-time digital video. In Proceedings of SPIE 4666: 149–160.
78. S. Li, X. Mou and Y. Cai. 2001. Improving security of a chaotic encryption approach. Physics Letters A 290(3-4): 127–133.
79. Ashwaq T. Hashim, Janan A. Mahdi and Salma H. Abdullah. 2010. A proposed 512 bits RC6 encryption algorithm. IJCCCE 10(1).
80. Fouad Ramia and G.M.U. Hunar Qadir. December, 2006. RC6 implementation including key scheduling using FPGA. ECE 646 Project.
81. A. Mohamed, G. Zaibi and A. Kachouri. March, 2011. Implementation of RC5 and RC6 bolck ciphers on digital images. Eighth International Multi-Conference on Systems, Signals & Devices.
82. N. El-Fishawy and O.M. Abu Zaid. 2007. Quality of encryption measurement of bitmap images with RC6, MRC6, and Rijndael block cipher algorithms. International Journal of Network Security 5(3): 241–251.

83. E. Lawrey. 1997. The Suitability of OFDM as a Modulation Technique for Wireless Telecommunications, with a CDMA comparison. Bachelor Thesis, James Cook University.
84. O.M. Enerstvedt. 2017. Analysis of privacy and data protection principles. pp. 307–394. *In*: Aviation Security, Privacy, Data Protection and Other Human Rights: Technologies and Legal Principles, Springer, Cham.
85. X. Zheng. 2017. The application of information security encryption technology in military data system management. pp. 423–428. *In*: International Conference on Man-Machine-Environment System Engineering, Springer, Singapore.
86. V. Madaan, D. Sethi, P. Agrawal, L. Jain and R. Kaur. 2017. Public network security by bluffing the intruders through encryption over encryption using public key cryptography method. pp. 249–257. *In*: Advanced Informatics for Computing Research, Springer, Singapore.
87. V.B. Durdi, P.T. Kulkarni and K.L. Sudha. 2017. Selective encryption framework for secure multimedia transmission over wireless multimedia sensor networks. pp. 469–480. *In*: Proceedings of the International Conference on Data Engineering and Communication Technology, Springer, Singapore.
88. K. Dhote and G.M. Asutkar. 2016. Enhancement in the performance of routing protocols for wireless communication using clustering, encryption, and cryptography. pp. 547–558. *In*: Artificial Intelligence and Evolutionary Computations in Engineering Systems, Springer, New Delhi.
89. P. Winkler. 2003. Mathematical Puzzles: A Connoisseur's Collection. A.K. Peters/CRC Press.
90. O. Alpar. 2017. A new chaotic map with three isolated chaotic regions. Nonlinear Dynamics, pp. 903–912.
91. L. Liu and S. Miao. 2018. A new simple one-dimensional chaotic map and its application for image encryption. Multimedia Tools and Applications, Springer, pp. 1–18.
92. X. Chai, Z. Gan and M. Zhang. 2017. A fast chaos-based image encryption scheme with a novel plain image-related swapping block permutation and block diffusion. Multimedia Tools and Applications, Springer, pp. 15561–15585.
93. Harsh Kumar Verma and Ravindra Kumar Singh. March, 2012. Performance analysis of RC6, Twofish and Rijndael block cipher algorithms. International Journal of Computer Applications 42(16).
94. Y. Liu, L.Y. Zhang, J. Wang, Y. Zhang and K.W. Wong. 2016. Chosen-plaintext attack of an image encryption scheme based on modified permutation–diffusion structure. Nonlinear Dynamics, pp. 2241–2250.
95. N. Varshney and K. Raghuwanshi. 2016. RC6 based data security and attack detection. pp. 3–10. *In*: Proceedings of First International Conference on Information and Communication Technology for Intelligent Systems, Springer International Publishing.
96. J. Wu, Z. Zhu and S. Guo. 2017. A quality model for evaluating encryption-as-a-service. pp. 557–569. *In*: International Conference on Security, Privacy and Anonymity in Computation, Communication and Storage, Springer, Cham.
97. Swiss Encryption Technology, MediCrypt, Modes of operation, pp. 1–4, http://www.mediacrypt.com/pdf/MC modes 1204pdf.
98. M. Hilmey, S. Elhalafwy and M. Eldin. 2009. Efficient transmission of chaotic and AES encrypted images with OFDM over an AWGN channel. pp. 353–358. *In*: Proceedings of the IEEE International Conference of Computer Engineering & Systems.
99. I. Eldokany, E. El-Rabaie, S. Elhalafawy, M. Eldin, M. Shahieen and Soliman N. El-Samie. 2015. Efficient transmission of encrypted images with OFDM in the presence of carrier frequency offset. Wireless Personal Communications, pp. 475–521.
100. M. Helmy, E. El-Rabaie, I. Eldokany and F. El-Samie. 2018. Chaotic encryption with different modes of operation based on Rubik's cube for efficient wireless communication. Multimedia Tools and Applications, Springer, pp. 1–25.
101. M. Helmy, E. El-Rabaie, I. Eldokany and F. El-Samie. 2017. 3-D Image Encryption Based on Rubik's Cube and RC6 Algorithm 3D Research pp. 38.

102. P. Tan and N.C. Beaulieu. 2005. Precise bit error probability analysis of DCT OFDM in the presence of carrier frequency offset on AWGN channels. In Proceedings of the IEEE Globcom, pp. 1429–1434.
103. E.P. Lawrey. 2001. Adaptive Techniques for Multiuser OFDM. Ph.D. Thesis, James Cook University.
104. W. Xiang, P. Gao and Q. Peng. 2015. Robust multiview three-dimensional video communications based on distributed video coding. IEEE Systems Journal, pp. 1–11.
105. O. Cagri, E. Erhan, C. Janko and K. Ahmet. 2016. Adaptive delivery of immersive 3D multi-view video over the Internet. Multimedia Tools and Application pp. 12431–12461.
106. Z. Huanqiang, W. Xiaolan, C. Canhui, C. Jing and Z. Yan. 2014. Fast multiview video coding using adaptive prediction structure and hierarchical mode decision. IEEE Transactions on Circuits and Systems for Video Technology, pp. 1566–1578.
107. W. El-Shafai, S. El-Rabaie, M.M. El-Halawany and F.E.A. El-Samie. 2017. Encoder-independent decoder-dependent depth-assisted error concealment algorithm for wireless 3D video communication. Multimedia Tools and Applications, Springer, pp. 1–28.
108. C.T.E.R. Hewage and M.G. Martini. 2013. Quality of experience for 3D video streaming. IEEE Communications Magazine, pp. 101–107.
109. Z. Liu, G. Cheung and Y. Ji. 2013. Optimizing distributed source coding for interactive multiview video streaming over lossy networks. IEEE Transactions on Circuits and Systems for Video Technology, pp. 1781–1794.
110. W. El-Shafai, S. El-Rabaie, M.M. El-Halawany and F.E.A. El-Samie. 2018. Performance evaluation of enhanced error correction algorithms for efficient wireless 3D video communication systems. International Journal of Communication Systems.
111. A. Abreu, P. Frossard and F. Pereira. 2015. Optimizing multiview video plus depth prediction structures for interactive multiview video streaming. IEEE Journal of Selected Topics in Signal Processing, pp. 487–500.
112. W. El-Shafai, S. El-Rabaie, M.M. El-Halawany and F.E.A. El-Samie. 2017. Enhancement of wireless 3D video communication using color-plus-depth error restoration algorithms and bayesian kalman filtering. Wireless Personal Communications, pp. 245–268.
113. Q.M. Rajpoot and C.D. Jensen. 2014. Security and privacy in video surveillance: requirements and challenges. pp. 169–184. *In*: IFIP International Information Security Conference, Springer, Berlin, Heidelberg.
114. L. Hu, Y. Li, T. Li, H. Li and J. Chu. 2016. The efficiency improved scheme for secure access control of digital video distribution. Multimedia Tools and Applications, Springer, pp. 12645–12662.
115. A. Souyah and K.M. Faraoun. 2016. A review on different image encryption approaches. pp. 3–18. *In*: Modelling and Implementation of Complex Systems, Springer, International Publishing.
116. A. Hamid, M. Samir, A. El-Atty, E. El-Hennawy, H. El Shenawy, A. Alshebeili and F. El Samie. 2014. On the performance of FFT/DWT/DCT-based OFDM systems with chaotic interleaving and channel estimation algorithms. Wireless Personal Communications, pp. 1495–1510.
117. R.H. Alsisi and R.K. Rao. 2017. BER Comparison of constant envelope DCT and FFT based OFDM with phase modulation. pp. 959–968. *In*: International Conference on Network-Based Information Systems, Springer, Cham.
118. ISO/IEC JTC1/SC29/WG11. 2006. Common test conditions for multiview video coding. JVT-U207, Hangzhou, China.
119. http://wftp3.itu.int/av-arch/jvt-site/2009_01_Geneva/JVT-AD207.zip: Reference software for multiview video coding (mvc), Last accessed on 25/10/2016.
120. http://iphome.hhi.de/suehring/tml/: H.264/AVC codec, 28/09/2016.

Appendix A: MATLAB® Codes for Rubik's Cube Encryption Algorithm

Rubik's Cube + CBC-Chaotic Algorithm over OFDM communication system / Same faces.....................
%INITIAL PARAMETER...............................
%Transmitter.......................................

```
im=imread('cameraman.tif');
im=double(im);
nw=128;
y=ChaoticCBC(im,nw);
figure(1);
imshow(y);
%%%%%%%%%%%%%%%%%%%%%%%%%%%%%%%%%%%%%%%%%%%%%%%%%%%%%
xall=[y y y;
    y y y;
    y y y];

n = [64,64,64,64,64,64,64,64,64,64,64,64];

[pr,pc] = chaomat(n);
pim = chaoperm(xall,pr,pc,3,'forward');
%%%%%%%%%%%%%%%%%%%%%%%%%%%%%%%%%%%%%%%%%%%5
figure(2);
imshow(pim/255);
xxx=imcrop(pim/255,[0 0 256 256]);
imshow(xxx)

imwrite(xxx,'ChaoticCBC_cameraman_Rubix.tif','tif');
%%%%%%%%%%%%%%%%%%%%%%%%%%%%%%%%%%%%%%%%%%%%%%%%%%%%%
[M,N]=size(pim) ;
g1 =im2col(pim, [M,N], [M,N], 'distinct');
h1=dec2bin(double(g1));
[M1,N1]=size(h1) ;
z1=zeros (M1,N1) ;
for i=1:M1
for j=1:N1
```

```
z1(i,j)=str2num(h1(i,j));
end;
end;
[R1,T1]=size(z1) ;
zz1=reshape(z1,R1*T1, 1);

% The transmitted data.................................
zz = [zz1];
            trel =poly2trellis(7,[171 133]); % Trellis
            data1=zz;
            zz = convenc(data1,trel);
s1=length(zz);
 para =128;
nd=6;       %number of information OFDM symbol for one
            loop
ml=2;       %Modulation level:QPSK
sr=250000;  %symbol rate
 br=sr.*ml;  %Bit rat per carrier
Ipoint = 8; %Number of over samples
gilen=32;
flat=1;
fd=600;

    dctlen=para;
    noc=para;
    ofdm_length = para*nd*ml;  %Total no for one loop

%Dividing the image into blocks........................
nloops = ceil((length(zz))/ofdm_length );
new_data = nloops*ofdm_length ;
nzeros   = new_data  - length(zz);
input_data = [zz;zeros(nzeros,1)];
input_data2 = reshape(input_data ,ofdm_length ,nloops);

%Transmission ON FFT_OFDM..............................

   for ebno=[0,2,4,6,8,10];
demodata1 = zeros(ofdm_length ,nloops);
for jj = 1: nloops         % loop for columns
    serdata1 = input_data2(:,jj)';
    demodata = fft_fading_channel(serdata1,para,nd,ml,
    gilen,dctlen,sr,ebno, br);
   demodata1(:,jj) = demodata(:);      % the output of
   ofdm columns

end

%Received image........................................
 [Mr,Nr] = size(demodata1);
%  demodata2 = demodata1(:);
  yy      = reshape (demodata1,Mr*Nr,1); %
 part1   = yy(1:s1);
```

```
yy11=part1;
%%%%%%%%%%%%%%%%%%%%%%%%%%%%%%%%DECODING...........
 yy11 = vitdec(yy11',trel,1,'term','hard'); % Decode.

yy1=reshape(yy11,[R1,T1]);
for i=1:M1
for j=1:N1
zn1(i,j)=num2str(yy1(i,j));
end;
end;
hn1=bin2dec(zn1);
gn1=col2im(hn1, [M,N], [M,N], 'distinct');

%%%%%%%%%%%%%%%%%%%%%%%%%%%%%%%%%%%%%%%%%%%%%%%%%5
n = [64,64,64,64,64,64,64,64,64,64,64,64];
[pr,pc] = chaomat(n);
rim0 = chaoperm(gn1,pr,pc,3,'backward');
figure(22);
imshow(rim0/255);
rim=imcrop(rim0,[0 0 256 256]);
imshow(rim)

gn11=uint8(rim);


% restore the original image
y1=ChaoticCBCdec(gn11,nw);
% figure(2);
y1=double(y1);
rim1=medfilt2(y1);
output_image=(y1/255);
output_image1=(rim1/255);

%The Error between Trans.............................
 MSE1=sum(sum((double(im)/255-output_image).^2))/
 prod(size(im));
PSNR=10*log(1/MSE1)/log(10)
 MSE11=sum(sum((double(im)/255-output_image1).^2))/
 prod(size(im));
PSNR1=10*log(1/MSE11)/log(10)
    end
%%%%%%%%%%%%%%%%%%%%%%%%%%%%%%%%%%%%%%%%%%%%%%%%%
%Transmission ON FFT_OFDM...........................

   for ebno=[0,2,4,6,8,10];

demodata1 = zeros(ofdm_length ,nloops);
for jj = 1: nloops          % loop for columns
    serdata1 = input_data2(:,jj)';
    demodata = dct_fading_channel(serdata1,para,nd,ml,
    gilen,dctlen,sr,ebno, br);
```

```
   demodata1(:,jj) = demodata(:);    % the output of
   ofdm columns

end

%Received image......................................
 [Mr,Nr] = size(demodata1);
%  demodata2 = demodata1(:);
  yy      = reshape (demodata1,Mr*Nr,1); %
 part1   = yy(1:s1);
yy11=part1;
%%%%%%%%%%%%%%%%%%%%%%%%%%%%%%%%%DECODING...........
 yy11 = vitdec(yy11',trel,1,'term','hard'); % Decode.

yy1=reshape(yy11,[R1,T1]);
for i=1:M1
for j=1:N1
zn1(i,j)=num2str(yy1(i,j));
end;
end;
hn1=bin2dec(zn1);
gn1=col2im(hn1, [M,N], [M,N], 'distinct');

%%%%%%%%%%%%%%%%%%%%%%%%%%%%%%%%%%%%%%%%%%%%%%%%%%%%%%%%5
n = [64,64,64,64,64,64,64,64,64,64,64,64];
[pr,pc] = chaomat(n);
rim0 = chaoperm(gn1,pr,pc,3,'backward');
figure(22);
imshow(rim0/255);
rim=imcrop(rim0,[0 0 256 256]);
imshow(rim)

gn11=uint8(rim);


% restore the original image
y1=ChaoticCBCdec(gn11,nw);
% figure(2);
y1=double(y1);
rim1=medfilt2(y1);
output_image=(y1/255);
output_image1=(rim1/255);

%The Error between Trans.............................
 MSE1=sum(sum((double(im)/255-output_image).^2))/
 prod(size(im));
PSNR=10*log(1/MSE1)/log(10)
 MSE11=sum(sum((double(im)/255-output_image1).^2))/
 prod(size(im));
PSNR1=10*log(1/MSE11)/log(10)
    end
%%%%%%%%%%%%%%%%%%%%%%%%%%%%%%%%%%%%%%%%%%%%%%%%%%%%%%%
```

```
%Transmission ON FFT_OFDM.............................
   for ebno=[0,2,4,6,8,10];

demodata1 = zeros(ofdm_length ,nloops);
for jj = 1: nloops          % loop for columns
    serdata1 = input_data2(:,jj)';
    demodata = dct_fading_channel(serdata1,para,nd,ml,
    gilen,dctlen,sr,ebno, br);
   demodata1(:,jj) = demodata(:);    % the output of
   ofdm columns

end

%Received image.......................................
 [Mr,Nr] = size(demodata1);
%  demodata2 = demodata1(:);
  yy      = reshape (demodata1,Mr*Nr,1); %
 part1   = yy(1:s1);
yy11=part1;
%%%%%%%%%%%%%%%%%%%%%%%%%%%%%%%%%DECODING...........
 yy11 = vitdec(yy11',trel,1,'term','hard'); % Decode.
yy1=reshape(yy11,[R1,T1]);
for i=1:M1
for j=1:N1
zn1(i,j)=num2str(yy1(i,j));
end;
end;
hn1=bin2dec(zn1);
gn1=col2im(hn1, [M,N], [M,N], 'distinct');

%%%%%%%%%%%%%%%%%%%%%%%%%%%%%%%%%%%%%%%%%%%%%%%%%%5
n = [64,64,64,64,64,64,64,64,64,64,64,64];
[pr,pc] = chaomat(n);
rim0 = chaoperm(gn1,pr,pc,3,'backward');
figure(22);
imshow(rim0/255);
rim=imcrop(rim0,[0 0 256 256]);
imshow(rim)
gn11=uint8(rim);
% restore the original image
y1=ChaoticCBCdec(gn11,nw);
% figure(2);
y1=double(y1);
rim1=medfilt2(y1);
output_image=(y1/255);
output_image1=(rim1/255);

%The Error between Trans..............................
 MSE1=sum(sum((double(im)/255-output_image).^2))/
 prod(size(im));
PSNR=10*log(1/MSE1)/log(10)
```

```
 MSE11=sum(sum((double(im)/255-output_image1).^2))/
 prod(size(im));
PSNR1=10*log(1/MSE11)/log(10);
    end
```

Rubik's Cube + CFB-Chaotic Algorithm over OFDM communication system / Same faces......................

```
%INITIAL PARAMETER...............................
%Transmitter..................................
 %INITIAL PARAMETER..............................
%Transmitter..................................

im=imread('cameraman.tif');
im=double(im);
nw=128;
y=ChaoticCFB(im,nw);
figure(1);
imshow(y);
%%%%%%%%%%%%%%%%%%%%%%%%%%%%%%%%%%%%%%%%%%%%%%%%%%%%%
xa11=[y y y;
    y y y;
    y y y];

n = [64,64,64,64,64,64,64,64,64,64,64,64];

[pr,pc] = chaomat(n);
pim = chaoperm(xa11,pr,pc,3,'forward');
%%%%%%%%%%%%%%%%%%%%%%%%%%%%%%%%%%%%%%%%5
figure(2);
imshow(pim/255);
xxx=imcrop(pim/255,[0 0 256 256]);
imshow(xxx)

imwrite(xxx,'ChaoticCFB_cameraman_Rubix.tif','tif');
%%%%%%%%%%%%%%%%%%%%%%%%%%%%%%%%%%%%%%%%%%%%%%%%%%%%%
[M,N]=size(pim) ;
g1 =im2col(pim, [M,N], [M,N], 'distinct');
h1=dec2bin(double(g1));
[M1,N1]=size(h1) ;
z1=zeros (M1,N1) ;
for i=1:M1
for j=1:N1
z1(i,j)=str2num(h1(i,j));
end;
end;
[R1,T1]=size(z1) ;
zz1=reshape(z1,R1*T1, 1);

% The transmitted data..............................
zz = [zz1];
            trel =poly2trellis(7,[171 133]); % Trellis
```

```
            data1=zz;
            zz = convenc(data1,trel);
s1=length(zz);
 para =128;
nd=6;       %number of information OFDM symbol for one
            loop
ml=2;       %Modulation level:QPSK
sr=250000;  %symbol rate
 br=sr.*ml;  %Bit rat per carrier
Ipoint = 8; %Number of over samples
gilen=32;
flat=1;
fd=100;

    dctlen=para;
    fftlen=para;
    noc=para;
    ofdm_length = para*nd*ml;   %Total no for one loop

%Dividing the image into blocks......................
nloops = ceil((length(zz))/ofdm_length );
new_data = nloops*ofdm_length ;
nzeros   = new_data  - length(zz);
input_data = [zz;zeros(nzeros,1)];
input_data2 = reshape(input_data ,ofdm_length ,nloops);

%Transmission ON FFT_OFDM.............................

   for ebno=[0,2,4,6,8,10];

demodata1 = zeros(ofdm_length ,nloops);
for jj = 1: nloops          % loop for columns
    serdata1 = input_data2(:,jj)';
    demodata = fft_channel_estimation_no_mapping_zg
    (serdata1,para,nd,ml,gilen,fftlen,sr,ebno,
    br,fd,flat);
   demodata1(:,jj) = demodata(:);   % the output of
   ofdm columns

end

%Received image.......................................
 [Mr,Nr] = size(demodata1);
%  demodata2 = demodata1(:);
  yy      = reshape (demodata1,Mr*Nr,1); %
 part1   = yy(1:s1);
yy11=part1;
%%%%%%%%%%%%%%%%%%%%%%%%%%%%%%%%DECODING..........
 yy11 = vitdec(yy11',trel,1,'term','hard'); % Decode.

yy1=reshape(yy11,[R1,T1]);
for i=1:M1
for j=1:N1
```

```
zn1(i,j)=num2str(yy1(i,j));
end;
end;
hn1=bin2dec(zn1);
gn1=col2im(hn1, [M,N], [M,N], 'distinct');

%%%%%%%%%%%%%%%%%%%%%%%%%%%%%%%%%%%%%%%%%%%%%%%%%%%%
n = [64,64,64,64,64,64,64,64,64,64,64,64];
[pr,pc] = chaomat(n);
rim0 = chaoperm(gn1,pr,pc,3,'backward');
figure(22);
imshow(rim0/255);
rim=imcrop(rim0,[0 0 256 256]);
imshow(rim)

gn11=uint8(rim);

% restore the original image
y1=ChaoticCFBdec(gn11,nw);
% figure(2);
y1=double(y1);
rim1=medfilt2(y1);
output_image=(y1/255);
output_image1=(rim1/255);

%The Error between Trans.............................
 MSE1=sum(sum((double(im)/255-output_image).^2))/
 prod(size(im));
PSNR=10*log(1/MSE1)/log(10)
 MSE11=sum(sum((double(im)/255-output_image1).^2))/
 prod(size(im));
PSNR1=10*log(1/MSE11)/log(10)
    end
%%%%%%%%%%%%%%%%%%%%%%%%%%%%%%%%%%%%%%%%%%%%%%%%%%%
%Transmission ON FFT_OFDM...........................

   for ebno=[0,2,4,6,8,10];

demodata1 = zeros(ofdm_length ,nloops);
for jj = 1: nloops          % loop for columns
    serdata1 = input_data2(:,jj)';
    demodata = dct_channel_estimation_no_mapping_zg
    (serdata1,para,nd,ml,gilen,fftlen,sr,ebno,
    br,fd,flat);
   demodata1(:,jj) = demodata(:);    % the output of
   ofdm columns

end

%Received image.....................................
 [Mr,Nr] = size(demodata1);
%  demodata2 = demodata1(:);
```

```
  yy      = reshape (demodata1,Mr*Nr,1); %
 part1   = yy(1:s1);
yy11=part1;
%%%%%%%%%%%%%%%%%%%%%%%%%%%%%%%%%DECODING...........
 yy11 = vitdec(yy11',trel,1,'term','hard'); % Decode.

yy1=reshape(yy11,[R1,T1]);
for i=1:M1
for j=1:N1
zn1(i,j)=num2str(yy1(i,j));
end;
end;
hn1=bin2dec(zn1);
gn1=col2im(hn1, [M,N], [M,N], 'distinct');

%%%%%%%%%%%%%%%%%%%%%%%%%%%%%%%%%%%%%%%%%%%%%%%%%%5
n = [64,64,64,64,64,64,64,64,64,64,64,64];
[pr,pc] = chaomat(n);
rim0 = chaoperm(gn1,pr,pc,3,'backward');
figure(22);
imshow(rim0/255);
rim=imcrop(rim0,[0 0 256 256]);
imshow(rim)

gn11=uint8(rim);

% restore the original image
y1=ChaoticCFBdec(gn11,nw);
% figure(2);
y1=double(y1);
rim1=medfilt2(y1);
output_image=(y1/255);
output_image1=(rim1/255);

%The Error between Trans............................
 MSE1=sum(sum((double(im)/255-output_image).^2))/
 prod(size(im));
PSNR=10*log(1/MSE1)/log(10)
 MSE11=sum(sum((double(im)/255-output_image1).^2))/
 prod(size(im));
PSNR1=10*log(1/MSE11)/log(10)
    end
%%%%%%%%%%%%%%%%%%%%%%%%%%%%%%%%%%%%%%%%%%%%%%%%%
%Transmission ON FFT_OFDM...........................

   for ebno=[0,2,4,6,8,10];

demodata1 = zeros(ofdm_length ,nloops);
for jj = 1: nloops          % loop for columns
    serdata1 = input_data2(:,jj)';
    demodata = dwt_channel_estimation_no_mapping_zg
```

```
    (serdata1,para,nd,ml,gilen,fftlen,sr,ebno,
    br,fd,flat);
   demodata1(:,jj) = demodata(:);    % the output of
   ofdm columns
end

%Received image.......................................
 [Mr,Nr] = size(demodata1);
%  demodata2 = demodata1(:);
  yy      = reshape (demodata1,Mr*Nr,1); %
 part1   = yy(1:s1);
yy11=part1;
%%%%%%%%%%%%%%%%%%%%%%%%%%%%%%%%%DECODING...........
 yy11 = vitdec(yy11',trel,1,'term','hard'); % Decode.

yy1=reshape(yy11,[R1,T1]);
for i=1:M1
for j=1:N1
zn1(i,j)=num2str(yy1(i,j));
end;
end;
hn1=bin2dec(zn1);
gn1=col2im(hn1, [M,N], [M,N], 'distinct');

%%%%%%%%%%%%%%%%%%%%%%%%%%%%%%%%%%%%%%%%%%%%%%%%%%5
n = [64,64,64,64,64,64,64,64,64,64,64,64];
[pr,pc] = chaomat(n);
rim0 = chaoperm(gn1,pr,pc,3,'backward');
figure(22);
imshow(rim0/255);
rim=imcrop(rim0,[0 0 256 256]);
imshow(rim)

gn11=uint8(rim);

% restore the original image
y1=ChaoticCFBdec(gn11,nw);
% figure(2);
y1=double(y1);
rim1=medfilt2(y1);
output_image=(y1/255);
output_image1=(rim1/255);

%The Error between Trans.............................
 MSE1=sum(sum((double(im)/255-output_image).^2))/
 prod(size(im));
PSNR=10*log(1/MSE1)/log(10)
 MSE11=sum(sum((double(im)/255-output_image1).^2))/
 prod(size(im));
PSNR1=10*log(1/MSE11)/log(10)
    end
```

Rubik's Cube + OFB-Chaotic Algorithm over OFDM communication system / Same faces....................
%INITIAL PARAMETER................................
%Transmitter.................................

```
im=imread('cameraman.tif');
im=double(im);
nw=128;
y=ChaoticOFB(im,nw);
figure(1);
imshow(y);
%%%%%%%%%%%%%%%%%%%%%%%%%%%%%%%%%%%%%%%%%%%%%%%%%%%%%%%
xall=[y y y;
    y y y;
    y y y];
n = [64,64,64,64,64,64,64,64,64,64,64,64];
[pr,pc] = chaomat(n);
pim = chaoperm(xall,pr,pc,3,'forward');
%%%%%%%%%%%%%%%%%%%%%%%%%%%%%%%%%%%%%%%%%%%%%5
figure(2);
imshow(pim/255);
xxx=imcrop(pim/255,[0 0 256 256]);
imshow(xxx)

imwrite(xxx,'ChaoticCBC_cameraman_Rubix.tif','tif');
%%%%%%%%%%%%%%%%%%%%%%%%%%%%%%%%%%%%%%%%%%%%%%%%%%%%%%%
[M,N]=size(pim) ;
g1 =im2col(pim, [M,N], [M,N], 'distinct');
h1=dec2bin(double(g1));
[M1,N1]=size(h1) ;
z1=zeros (M1,N1) ;
for i=1:M1
for j=1:N1
z1(i,j)=str2num(h1(i,j));
end;
end;
[R1,T1]=size(z1) ;
zz1=reshape(z1,R1*T1, 1);

% The transmitted data...............................
zz = [zz1];
            trel =poly2trellis(7,[171 133]); % Trellis
            data1=zz;
            zz = convenc(data1,trel);
s1=length(zz);
 para =128;
nd=6;       %number of information OFDM symbol for one
            loop
ml=2;       %Modulation level:QPSK
```

```
sr=250000;  %symbol rate
 br=sr.*ml;  %Bit rat per carrier
Ipoint = 8; %Number of over samples
gilen=32;
flat=1;
fd=100;

    dctlen=para;
    fftlen=para;
    noc=para;
    ofdm_length = para*nd*ml;  %Total no for one loop

%Dividing the image into blocks.......................
nloops = ceil((length(zz))/ofdm_length );
new_data = nloops*ofdm_length ;
nzeros   = new_data  - length(zz);
input_data = [zz;zeros(nzeros,1)];
input_data2 = reshape(input_data ,ofdm_length ,nloops);

%Transmission ON FFT_OFDM.............................

   for ebno=[0,2,4,6,8,10];

demodata1 = zeros(ofdm_length ,nloops);
for jj = 1: nloops         % loop for columns
    serdata1 = input_data2(:,jj)';
    demodata = fft_channel_estimation_no_mapping_zg
    (serdata1,para,nd,ml,gilen,fftlen,sr,ebno,
    br,fd,flat);
  demodata1(:,jj) = demodata(:);   % the output of
  ofdm columns

end

%Received image.....................................
 [Mr,Nr] = size(demodata1);
%  demodata2 = demodata1(:);
  yy      = reshape (demodata1,Mr*Nr,1); %
 part1   = yy(1:s1);
yy11=part1;
%%%%%%%%%%%%%%%%%%%%%%%%%%%%%%%%%%DECODING..........
 yy11 = vitdec(yy11',trel,1,'term','hard'); % Decode.

yy1=reshape(yy11,[R1,T1]);
for i=1:M1
for j=1:N1
zn1(i,j)=num2str(yy1(i,j));
end;
end;
hn1=bin2dec(zn1);
gn1=col2im(hn1, [M,N], [M,N], 'distinct');

%%%%%%%%%%%%%%%%%%%%%%%%%%%%%%%%%%%%%%%%%%%%%%%%%%%%5
```

```
n = [64,64,64,64,64,64,64,64,64,64,64,64];
[pr,pc] = chaomat(n);
rim0 = chaoperm(gn1,pr,pc,3,'backward');
figure(22);
imshow(rim0/255);
rim=imcrop(rim0,[0 0 256 256]);
imshow(rim)

gn11=uint8(rim);


% restore the original image
y1=ChaoticOFBdec(gn11,nw);
% figure(2);
y1=double(y1);
rim1=medfilt2(y1);
output_image=(y1/255);
output_image1=(rim1/255);

%The Error between Trans..............................
 MSE1=sum(sum((double(im)/255-output_image).^2))/
 prod(size(im));
PSNR=10*log(1/MSE1)/log(10)
 MSE11=sum(sum((double(im)/255-output_image1).^2))/
 prod(size(im));
PSNR1=10*log(1/MSE11)/log(10)
    end
%%%%%%%%%%%%%%%%%%%%%%%%%%%%%%%%%%%%%%%%%%%%%%%%%%%%
%Transmission ON FFT_OFDM...........................

   for ebno=[0,2,4,6,8,10];

demodata1 = zeros(ofdm_length ,nloops);
for jj = 1: nloops          % loop for columns
    serdata1 = input_data2(:,jj)';
    demodata = dct_channel_estimation_no_mapping_zg
    (serdata1,para,nd,ml,gilen,fftlen,sr,ebno,
    br,fd,flat);
   demodata1(:,jj) = demodata(:);    % the output of
   ofdm columns

end

%Received image....................................
 [Mr,Nr] = size(demodata1);
%  demodata2 = demodata1(:);
  yy      = reshape (demodata1,Mr*Nr,1); %
 part1   = yy(1:s1);
yy11=part1;
%%%%%%%%%%%%%%%%%%%%%%%%%%%%%%%%%%DECODING..........
 yy11 = vitdec(yy11',trel,1,'term','hard'); % Decode.
```

```
yy1=reshape(yy11,[R1,T1]);
for i=1:M1
for j=1:N1
zn1(i,j)=num2str(yy1(i,j));
end;
end;
hn1=bin2dec(zn1);
gn1=col2im(hn1, [M,N], [M,N], 'distinct');

%%%%%%%%%%%%%%%%%%%%%%%%%%%%%%%%%%%%%%%%%%%%%%%%%%%%%%%%%%%%%%%%%%%%%%%%%%%%%%%%%%%%%%%%%%%%%%%%%%5
n = [64,64,64,64,64,64,64,64,64,64,64,64];
[pr,pc] = chaomat(n);
rim0 = chaoperm(gn1,pr,pc,3,'backward');
figure(22);
imshow(rim0/255);
rim=imcrop(rim0,[0 0 256 256]);
imshow(rim)

gn11=uint8(rim);


% restore the original image
y1=ChaoticOFBdec(gn11,nw);
% figure(2);
y1=double(y1);
rim1=medfilt2(y1);
output_image=(y1/255);
output_image1=(rim1/255);

%The Error between Trans..............................
 MSE1=sum(sum((double(im)/255-output_image).^2))/
 prod(size(im));
PSNR=10*log(1/MSE1)/log(10)
 MSE11=sum(sum((double(im)/255-output_image1).^2))/
 prod(size(im));
PSNR1=10*log(1/MSE11)/log(10)
    end
%%%%%%%%%%%%%%%%%%%%%%%%%%%%%%%%%%%%%%%%%%%%%%%%%%%%%%%%%%%%%%%%%%%%%%%%%%%%%%%%%%%%%%%%%%%%%%%%%%%%
%Transmission ON FFT_OFDM.............................

   for ebno=[0,2,4,6,8,10];

demodata1 = zeros(ofdm_length ,nloops);
for jj = 1: nloops          % loop for columns
    serdata1 = input_data2(:,jj)';
    demodata = dwt_channel_estimation_no_mapping_zg
    (serdata1,para,nd,ml,gilen,fftlen,sr,ebno,
    br,fd,flat);
   demodata1(:,jj) = demodata(:);   % the output of
   ofdm columns

end
```

```
%Received image......................................
 [Mr,Nr] = size(demodata1);
%  demodata2 = demodata1(:);
  yy      = reshape (demodata1,Mr*Nr,1); %
 part1    = yy(1:s1);
yy11=part1;
%%%%%%%%%%%%%%%%%%%%%%%%%%%%%%%%%DECODING...........
 yy11 = vitdec(yy11',trel,1,'term','hard'); % Decode.

yy1=reshape(yy11,[R1,T1]);
for i=1:M1
for j=1:N1
zn1(i,j)=num2str(yy1(i,j));
end;
end;
hn1=bin2dec(zn1);
gn1=col2im(hn1, [M,N], [M,N], 'distinct');

%%%%%%%%%%%%%%%%%%%%%%%%%%%%%%%%%%%%%%%%%%%%%%%%%%%%5
n = [64,64,64,64,64,64,64,64,64,64,64,64];
[pr,pc] = chaomat(n);
rim0 = chaoperm(gn1,pr,pc,3,'backward');
figure(22);
imshow(rim0/255);
rim=imcrop(rim0,[0 0 256 256]);
imshow(rim)

gn11=uint8(rim);

% restore the original image
y1=ChaoticOFBdec(gn11,nw);
% figure(2);
y1=double(y1);
rim1=medfilt2(y1);
output_image=(y1/255);
output_image1=(rim1/255);

%The Error between Trans..............................
 MSE1=sum(sum((double(im)/255-output_image).^2))/
 prod(size(im));
PSNR=10*log(1/MSE1)/log(10)
 MSE11=sum(sum((double(im)/255-output_image1).^2))/
 prod(size(im));
PSNR1=10*log(1/MSE11)/log(10)
    end
```

Appendix B: MATLAB® Codes for Encrypted Image Communication

```
%Image with no Encryption over OFDM system + AWGN.....
%Transmitter.........................................
%Data Generation.....................................

f=imread ('cameraman.tif');
[M,N]=size(f) ;
g1 =im2col(f, [M,N], [M,N], 'distinct');
h1=dec2bin(double(g1));
[M1,N1]=size(h1) ;
z1=zeros (M1,N1) ;
for i=1:M1
for j=1:N1
z1(i,j)=str2num(h1(i,j));
end;
end;
[R1,T1]=size(z1) ;
zz1=reshape(z1,R1*T1, 1);
% The transmitted data...............................
zz = [zz1];
            trel =poly2trellis(7,[171 133]); % Trellis
            data1=zz;
            zz = convenc(data1,trel);
s1=length(zz);
 para =128;
nd=6;       %number of information OFDM symbol for one
            loop
ml=2;       %Modulation level:QPSK
sr=250000;  %symbol rate
 br=sr.*ml;  %Bit rat per carrier
Ipoint = 8; %Number of over samples
gilen=32;
flat=1;
fd=600;
    fftlen=para;
```

```
    noc=para;
    ofdm_length = para*nd*ml;   %Total no for one loop

%Dividing the image into blocks......................
nloops = ceil((length(zz))/ofdm_length );
new_data = nloops*ofdm_length ;
nzeros   = new_data  - length(zz);
input_data = [zz;zeros(nzeros,1)];
input_data2 = reshape(input_data ,ofdm_length ,nloops);

%Transmission ON FFT_OFDM............................

   for ebno=[0,2,4,6,8,10];

demodata1 = zeros(ofdm_length ,nloops);
for jj = 1: nloops          % loop for columns
    serdata1 = input_data2(:,jj)';
    demodata = fft_channel(serdata1,para,nd,ml,gilen,f
    ftlen,sr,ebno, br);
   demodata1(:,jj) = demodata(:);   % the output of
   ofdm columns

end

%Received image.......................................
 [Mr,Nr] = size(demodata1);
%  demodata2 = demodata1(:);
  yy      = reshape (demodata1,Mr*Nr,1); %
 part1   = yy(1:s1);
yy11=part1;
%%%%%%%%%%%%%%%%%%%%%%%%%%%%%%%%%%DECODING..........
 yy11 = vitdec(yy11',trel,1,'term','hard'); % Decode.

yy1=reshape(yy11,[R1,T1]);
for i=1:M1
for j=1:N1
zn1(i,j)=num2str(yy1(i,j));
end;
end;
hn1=bin2dec(zn1);
gn1=col2im(hn1, [M,N], [M,N], 'distinct');

 %The Error between Trans.............................
   output_image = gn1/255;

output_image1=medfilt2( output_image);

%The Error between Trans..............................
 MSE1=sum(sum((double(f)/255-output_image).^2))/
 prod(size(f));
PSNR=10*log(1/MSE1)/log(10)
 MSE11=sum(sum((double(f)/255-output_image1).^2))/
```

```
 prod(size(f));
PSNR1=10*log(1/MSE11)/log(10);

   end

%Transmission ON DCT_OFDM..............................

   for ebno=[0,2,4,6,8,10];

demodata1 = zeros(ofdm_length ,nloops);
for jj = 1: nloops          % loop for columns
    serdata1 = input_data2(:,jj)';
    demodata = dct_channel(serdata1,para,nd,ml,gilen,f
    ftlen,sr,ebno, br);
   demodata1(:,jj) = demodata(:);   % the output of
   ofdm columns

end

%Received image.......................................
 [Mr,Nr] = size(demodata1);
%  demodata2 = demodata1(:);
  yy      = reshape (demodata1,Mr*Nr,1); %
 part1   = yy(1:s1);
yy11=part1;
%%%%%%%%%%%%%%%%%%%%%%%%%%%%%%%%%%DECODING...........
 yy11 = vitdec(yy11',trel,1,'term','hard'); % Decode.

yy1=reshape(yy11,[R1,T1]);
for i=1:M1
for j=1:N1
zn1(i,j)=num2str(yy1(i,j));
end;
end;
hn1=bin2dec(zn1);
gn1=col2im(hn1, [M,N], [M,N], 'distinct');
 %The Error between Trans..............................
   output_image = gn1/255;

output_image1=medfilt2( output_image);

%The Error between Trans...............................
 MSE1=sum(sum((double(f)/255-output_image).^2))/
 prod(size(f));
PSNR=10*log(1/MSE1)/log(10)
 MSE11=sum(sum((double(f)/255-output_image1).^2))/
 prod(size(f));
PSNR1=10*log(1/MSE11)/log(10);

   end
%Transmission ON DWT_OFDM..............................

   for ebno=[0,2,4,6,8,10];
```

```
demodata1 = zeros(ofdm_length ,nloops);
for jj = 1: nloops          % loop for columns
    serdata1 = input_data2(:,jj)';
    demodata = dwt_channel(serdata1,para,nd,ml,gilen,f
    ftlen,sr,ebno, br);
   demodata1(:,jj) = demodata(:);    % the output of
   ofdm columns

end

%Received image.......................................
 [Mr,Nr] = size(demodata1);
%  demodata2 = demodata1(:);
  yy       = reshape (demodata1,Mr*Nr,1); %
 part1    = yy(1:s1);
yy11=part1;
%%%%%%%%%%%%%%%%%%%%%%%%%%%%%%%%%%DECODING...........
 yy11 = vitdec(yy11',trel,1,'term','hard'); % Decode.

yy1=reshape(yy11,[R1,T1]);
for i=1:M1
for j=1:N1
zn1(i,j)=num2str(yy1(i,j));
end;
end;
hn1=bin2dec(zn1);
gn1=col2im(hn1, [M,N], [M,N], 'distinct');
 %The Error between Trans..............................
   output_image = gn1/255;

output_image1=medfilt2( output_image);

%The Error between Trans...............................
 MSE1=sum(sum((double(f)/255-output_image).^2))/
 prod(size(f));
PSNR=10*log(1/MSE1)/log(10)
 MSE11=sum(sum((double(f)/255-output_image1).^2))/
 prod(size(f));
PSNR1=10*log(1/MSE11)/log(10);

   end


%Original clipping + companding + AWGN.................
%Transmitter...........................................
%Data Generation.......................................

f=imread ('cameraman.tif' );
[M,N]=size(f) ;
g1 =im2col(f, [M,N], [M,N], 'distinct');
h1=dec2bin(double(g1));
[M1,N1]=size(h1) ;
```

```
z1=zeros (M1,N1) ;
for i=1:M1
for j=1:N1
z1(i,j)=str2num(h1(i,j));
end;
end;
[R1,T1]=size(z1) ;
zz1=reshape(z1,R1*T1, 1);
% The transmitted data................................
zz = [zz1];

%ENCODING..........................................
            trel =poly2trellis(7,[171 133]); % Trellis
            data1=zz;
            zz = convenc(data1,trel);
s1=length(zz);
 para =128;
nd=6;       %number of information OFDM symbol for one
            loop
ml=2;       %Modulation level:QPSK
sr=250000;  %symbol rate
 br=sr.*ml;  %Bit rat per carrier
Ipoint = 8; %Number of over samples
gilen=32;
flat=1;
fd=600;

    dctlen=para;
    noc=para;
    ofdm_length = para*nd*ml;   %Total no for one loop

%Dividing the image into blocks......................
nloops = ceil((length(zz))/ofdm_length );
new_data = nloops*ofdm_length ;
nzeros   = new_data  - length(zz);
input_data = [zz;zeros(nzeros,1)];
input_data2 = reshape(input_data ,ofdm_length ,nloops);

%Transmission ON FFT_OFDM...........................

   for ebno=[0,2,4,6,8,10];

demodata1 = zeros(ofdm_length ,nloops);
for jj = 1: nloops          % loop for columns
    serdata1 = input_data2(:,jj)';
    demodata = fft_channel_clipping_companding(serdata
    1,para,nd,ml,gilen,dctlen,sr,ebno, br);
   demodata1(:,jj) = demodata(:);   % the output of
   ofdm columns

end
```

```
%Received image.....................................
 [Mr,Nr] = size(demodata1);
%  demodata2 = demodata1(:);
  yy      = reshape (demodata1,Mr*Nr,1); %
 part1   = yy(1:s1);
yy11=part1;
%DECODING............................................
 yy11 = vitdec(yy11',trel,1,'term','hard'); % Decode.

yy1=reshape(yy11,[R1,T1]);
for i=1:M1
for j=1:N1
zn1(i,j)=num2str(yy1(i,j));
end;
end;
hn1=bin2dec(zn1);
gn1=col2im(hn1, [M,N], [M,N], 'distinct');
 %The Error between Trans.............................
   output_image = gn1/255;

output_image1=medfilt2( output_image);

%The Error between Trans..............................
 MSE1=sum(sum((double(f)/255-output_image).^2))/
 prod(size(f));
PSNR=10*log(1/MSE1)/log(10)
 MSE11=sum(sum((double(f)/255-output_image1).^2))/
 prod(size(f));
PSNR1=10*log(1/MSE11)/log(10);

   end

   %Transmission ON DCT_OFDM.........................

   for ebno=[0,2,4,6,8,10];

demodata1 = zeros(ofdm_length ,nloops);
for jj = 1: nloops           % loop for columns
    serdata1 = input_data2(:,jj)';
    demodata = dct_channel_clipping_companding(serdata
    1,para,nd,ml,gilen,dctlen,sr,ebno, br);
   demodata1(:,jj) = demodata(:);   % the output of
   ofdm columns

end

%Received image.....................................
 [Mr,Nr] = size(demodata1);
%  demodata2 = demodata1(:);
  yy      = reshape (demodata1,Mr*Nr,1); %
 part1   = yy(1:s1);
yy11=part1;
```

```
%DECODING.........................................
 yy11 = vitdec(yy11',trel,1,'term','hard'); % Decode.

yy1=reshape(yy11,[R1,T1]);
for i=1:M1
for j=1:N1
zn1(i,j)=num2str(yy1(i,j));
end;
end;
hn1=bin2dec(zn1);
gn1=col2im(hn1, [M,N], [M,N], 'distinct');
 %The Error between Trans..............................
   output_image = gn1/255;

output_image1=medfilt2( output_image);

%The Error between Trans..............................
 MSE1=sum(sum((double(f)/255-output_image).^2))/
 prod(size(f));
PSNR=10*log(1/MSE1)/log(10)
 MSE11=sum(sum((double(f)/255-output_image1).^2))/
 prod(size(f));
PSNR1=10*log(1/MSE11)/log(10);

   end

   %Transmission ON DWT_OFDM...........................

   for ebno=[0,2,4,6,8,10];

demodata1 = zeros(ofdm_length ,nloops);
for jj = 1: nloops         % loop for columns
    serdata1 = input_data2(:,jj)';
    demodata = dwt_channel_clipping_companding(serdata
    1,para,nd,ml,gilen,dctlen,sr,ebno, br);
   demodata1(:,jj) = demodata(:);    % the output of
   ofdm columns

end

%Received image...................................
 [Mr,Nr] = size(demodata1);
%  demodata2 = demodata1(:);
  yy      = reshape (demodata1,Mr*Nr,1); %
 part1   = yy(1:s1);
yy11=part1;
%DECODING.........................................
 yy11 = vitdec(yy11',trel,1,'term','hard'); % Decode.

yy1=reshape(yy11,[R1,T1]);
for i=1:M1
for j=1:N1
```

```
zn1(i,j)=num2str(yy1(i,j));
end;
end;
hn1=bin2dec(zn1);
gn1=col2im(hn1, [M,N], [M,N], 'distinct');
 %The Error between Trans.............................
   output_image = gn1/255;

output_image1=medfilt2( output_image);

%The Error between Trans.............................
 MSE1=sum(sum((double(f)/255-output_image).^2))/
 prod(size(f));
PSNR=10*log(1/MSE1)/log(10)
 MSE11=sum(sum((double(f)/255-output_image1).^2))/
 prod(size(f));
PSNR1=10*log(1/MSE11)/log(10);

   end
```

```
%Image without Encryption + clipping over estimated
channel.........................................
%Transmitter.....................................
%Data Generation.................................

f=imread ('cameraman.tif' );
[M,N]=size(f) ;
g1 =im2col(f, [M,N], [M,N], 'distinct');
h1=dec2bin(double(g1));
[M1,N1]=size(h1) ;
z1=zeros (M1,N1) ;
for i=1:M1
for j=1:N1
z1(i,j)=str2num(h1(i,j));
end;
end;
[R1,T1]=size(z1) ;
zz1=reshape(z1,R1*T1, 1);
% The transmitted data............................
zz = [zz1];

%ENCODING.........................................
            trel =poly2trellis(7,[171 133]); % Trellis
            data1=zz;
            zz = convenc(data1,trel);
s1=length(zz);
 para =128;
nd=6;       %number of information OFDM symbol for one
            loop
m1=2;       %Modulation level:QPSK
sr=250000;  %symbol rate
```

```
 br=sr.*m1;  %Bit rat per carrier
Ipoint = 8; %Number of over samples
gilen=32;
flat=1;
fd=600;

    fftlen=para;
    noc=para;
    ofdm_length = para*nd*m1;   %Total no for one loop

%Dividing the image into blocks......................
nloops = ceil((length(zz))/ofdm_length );
new_data = nloops*ofdm_length ;
nzeros   = new_data  - length(zz);
input_data = [zz;zeros(nzeros,1)];
input_data2 = reshape(input_data ,ofdm_length ,nloops);

%Transmission ON FFT_OFDM............................

   for ebno=[0,2,4,6,8,10];

demodata1 = zeros(ofdm_length ,nloops);
for jj = 1: nloops          % loop for columns
    serdata = input_data2(:,jj)';
    demodata = fft_channel_estimation_no_mapping_cl
    ipping(serdata,para,nd,m1,gilen,fftlen,sr,ebno,
    br,fd,flat);
   demodata1(:,jj) = demodata(:);   % the output of
   ofdm columns

end

%Received image......................................
 [Mr,Nr] = size(demodata1);
%  demodata2 = demodata1(:);
  yy      = reshape (demodata1,Mr*Nr,1); %
 part1   = yy(1:s1);
yy11=part1;
%DECODING............................................
 yy11 = vitdec(yy11',trel,1,'term','hard'); % Decode.

yy1=reshape(yy11,[R1,T1]);
for i=1:M1
for j=1:N1
zn1(i,j)=num2str(yy1(i,j));
end;
end;
hn1=bin2dec(zn1);
gn1=col2im(hn1, [M,N], [M,N], 'distinct');
 %The Error between Trans.............................
   output_image = gn1/255;
```

```
output_image1=medfilt2( output_image);

%The Error between Trans...............................
 MSE1=sum(sum((double(f)/255-output_image).^2))/
 prod(size(f));
PSNR=10*log(1/MSE1)/log(10)
 MSE11=sum(sum((double(f)/255-output_image1).^2))/
 prod(size(f));
PSNR1=10*log(1/MSE11)/log(10);

   end

   %Transmission ON DCT_OFDM..........................

   for ebno=[0,2,4,6,8,10];

demodata1 = zeros(ofdm_length ,nloops);
for jj = 1: nloops          % loop for columns
    serdata = input_data2(:,jj)';
    demodata =  dct_channel_estimation_no_mapping_cl
    ipping(serdata,para,nd,m1,gilen,fftlen,sr,ebno,
    br,fd,flat);
   demodata1(:,jj) = demodata(:);   % the output of
   ofdm columns

end

%Received image.......................................
 [Mr,Nr] = size(demodata1);
%  demodata2 = demodata1(:);
  yy      = reshape (demodata1,Mr*Nr,1); %
 part1   = yy(1:s1);
yy11=part1;
%DECODING.............................................
 yy11 = vitdec(yy11',trel,1,'term','hard'); % Decode.

yy1=reshape(yy11,[R1,T1]);
for i=1:M1
for j=1:N1
zn1(i,j)=num2str(yy1(i,j));
end;
end;
hn1=bin2dec(zn1);
gn1=col2im(hn1, [M,N], [M,N], 'distinct');
 %The Error between Trans............................
   output_image = gn1/255;

output_image1=medfilt2( output_image);

%The Error between Trans..............................
 MSE1=sum(sum((double(f)/255-output_image).^2))/
 prod(size(f));
```

```
PSNR=10*log(1/MSE1)/log(10)
 MSE11=sum(sum((double(f)/255-output_image1).^2))/
 prod(size(f));
PSNR1=10*log(1/MSE11)/log(10);

   end

   %Transmission ON DWT_OFDM..........................

   for  ebno=[0,2,4,6,8,10];

demodata1 = zeros(ofdm_length ,nloops);
for jj = 1: nloops          % loop for columns
    serdata = input_data2(:,jj)';
    demodata = dwt_channel_estimation_no_mapping_cl
    ipping(serdata,para,nd,m1,gilen,fftlen,sr,ebno,
    br,fd,flat);
   demodata1(:,jj) = demodata(:);   % the output of
   ofdm columns

end

%Received image....................................
 [Mr,Nr] = size(demodata1);
%  demodata2 = demodata1(:);
  yy      = reshape (demodata1,Mr*Nr,1); %
 part1   = yy(1:s1);
yy11=part1;
%DECODING..........................................
 yy11 = vitdec(yy11',trel,1,'term','hard'); % Decode.

yy1=reshape(yy11,[R1,T1]);
for i=1:M1
for j=1:N1
zn1(i,j)=num2str(yy1(i,j));
end;
end;
hn1=bin2dec(zn1);
gn1=col2im(hn1, [M,N], [M,N], 'distinct');
 %The Error between Trans...........................
   output_image = gn1/255;

output_image1=medfilt2( output_image);

%The Error between Trans............................
 MSE1=sum(sum((double(f)/255-output_image).^2))/
 prod(size(f));
PSNR=10*log(1/MSE1)/log(10)
 MSE11=sum(sum((double(f)/255-output_image1).^2))/
 prod(size(f));
PSNR1=10*log(1/MSE11)/log(10);
   end
```

```
%Image with no Encryption over OFDM system + offset +
cyclic prefix + AWGN..................................
%Transmitter..........................................
%Data Generation......................................
%Transmitter..........................................
%Data Generation......................................

f=imread ('cameraman.tif' );
[M,N]=size(f) ;
g1 =im2col(f, [M,N], [M,N], 'distinct');
h1=dec2bin(double(g1));
[M1,N1]=size(h1) ;
z1=zeros (M1,N1) ;
for i=1:M1
for j=1:N1
z1(i,j)=str2num(h1(i,j));
end;
end;
[R1,T1]=size(z1) ;
zz1=reshape(z1,R1*T1, 1);
% The transmitted data................................
zz = [zz1];
            trel =poly2trellis(7,[171 133]); % Trellis
            data1=zz;
            zz = convenc(data1,trel);
s1=length(zz);
 para =128;
nd=6;       %number of information OFDM symbol for one
            loop
m1=2;       %Modulation level:QPSK
sr=250000;  %symbol rate
 br=sr.*ml;  %Bit rat per carrier
Ipoint = 8; %Number of over samples
gilen=32;
flat=1;
fd=600;
epsilon=0.1;
    fftlen=para;
    noc=para;
    ofdm_length = para*nd*ml;   %Total no for one loop

%Dividing the image into blocks.......................
nloops = ceil((length(zz))/ofdm_length );
new_data = nloops*ofdm_length ;
nzeros   = new_data  - length(zz);
input_data = [zz;zeros(nzeros,1)];
input_data2 = reshape(input_data ,ofdm_length ,nloops);

%Transmission ON FFT_OFDM.............................

   for ebno=[0,2,4,6,8,10];
```

```
demodata1 = zeros(ofdm_length ,nloops);
for jj = 1: nloops          % loop for columns
    serdata = input_data2(:,jj)';
    demodata = fft_channel_offset_g(serdata,para,nd,m1
    ,gilen,fftlen,sr,ebno, br,epsilon);
   demodata1(:,jj) = demodata(:);   % the output of
   ofdm columns

end

%Received image..................................
 [Mr,Nr] = size(demodata1);
%  demodata2 = demodata1(:);
  yy      = reshape (demodata1,Mr*Nr,1); %
 part1   = yy(1:s1);
yy11=part1;
%%%%%%%%%%%%%%%%%%%%%%%%%%%%%%%%%%%DECODING...........
 yy11 = vitdec(yy11',trel,1,'term','hard'); % Decode.

yy1=reshape(yy11,[R1,T1]);
for i=1:M1
for j=1:N1
zn1(i,j)=num2str(yy1(i,j));
end;
end;
hn1=bin2dec(zn1);
gn1=col2im(hn1, [M,N], [M,N], 'distinct');
 %The Error between Trans.............................
   output_image = gn1/255;

output_image1=medfilt2( output_image);

%The Error between Trans.............................
 MSE1=sum(sum((double(f)/255-output_image).^2))/
 prod(size(f));
PSNR=10*log(1/MSE1)/log(10)
 MSE11=sum(sum((double(f)/255-output_image1).^2))/
 prod(size(f));
PSNR1=10*log(1/MSE11)/log(10);

   end

%Transmission ON DCT_OFDM............................

   for ebno=[0,2,4,6,8,10];

demodata1 = zeros(ofdm_length ,nloops);
for jj = 1: nloops          % loop for columns
    serdata1 = input_data2(:,jj)';
    demodata = dct_channel_offset_g(serdata1,para,nd,m
    1,gilen,fftlen,sr,ebno, br,epsilon);
```

```
  demodata1(:,jj) = demodata(:);   % the output of
  ofdm columns
end

%Received image...................................
 [Mr,Nr] = size(demodata1);
%  demodata2 = demodata1(:);
  yy      = reshape (demodata1,Mr*Nr,1); %
 part1   = yy(1:s1);
yy11=part1;
%%%%%%%%%%%%%%%%%%%%%%%%%%%%%%%%%%%%DECODING..........
 yy11 = vitdec(yy11',trel,1,'term','hard'); % Decode.

yy1=reshape(yy11,[R1,T1]);
for i=1:M1
for j=1:N1
zn1(i,j)=num2str(yy1(i,j));
end;
end;
hn1=bin2dec(zn1);
gn1=col2im(hn1, [M,N], [M,N], 'distinct');
 %The Error between Trans............................
   output_image = gn1/255;

output_image1=medfilt2( output_image);

%The Error between Tras.............................
 MSE1=sum(sum((double(f)/255-output_image).^2))/
 prod(size(f));
PSNR=10*log(1/MSE1)/log(10)
 MSE11=sum(sum((double(f)/255-output_image1).^2))/
 prod(size(f));
PSNR1=10*log(1/MSE11)/log(10);

   end

%Transmission ON DWT_OFDM...........................

   for ebno=[0,2,4,6,8,10];

demodata1 = zeros(ofdm_length ,nloops);
for jj = 1: nloops          % loop for columns
    serdata1 = input_data2(:,jj)';
    demodata =dwt_channel_offset_g(serdata1,para,nd,m1
    ,gilen,fftlen,sr,ebno, br,epsilon);
  demodata1(:,jj) = demodata(:);   % the output of
  ofdm columns

end

%Received image...................................
 [Mr,Nr] = size(demodata1);
%  demodata2 = demodata1(:);
```

```
  yy      = reshape (demodata1,Mr*Nr,1); %
 part1   = yy(1:s1);
yy11=part1;
%%%%%%%%%%%%%%%%%%%%%%%%%%%%%%%%%%DECODING...........
 yy11 = vitdec(yy11',trel,1,'term','hard'); % Decode.

yy1=reshape(yy11,[R1,T1]);
for i=1:M1
for j=1:N1
zn1(i,j)=num2str(yy1(i,j));
end;
end;
hn1=bin2dec(zn1);
gn1=col2im(hn1, [M,N], [M,N], 'distinct');
 %The Error between Trans..............................
   output_image = gn1/255;

output_image1=medfilt2( output_image);

%The Error between Trans...............................
 MSE1=sum(sum((double(f)/255-output_image).^2))/
 prod(size(f));
PSNR=10*log(1/MSE1)/log(10)
 MSE11=sum(sum((double(f)/255-output_image1).^2))/
 prod(size(f));
PSNR1=10*log(1/MSE11)/log(10);

   end
```

```
%Image with no Encryption over OFDM system + offset +
zero padding + AWGN....................................
%Transmitter...........................................
%Data Generation.......................................

f=imread ('cameraman.tif' );
[M,N]=size(f) ;
g1 -im2col(f, [M,N], [M,N], 'distinct');
h1=dec2bin(double(g1));
[M1,N1]=size(h1) ;
z1=zeros (M1,N1) ;
for i=1:M1
for j=1:N1
z1(i,j)=str2num(h1(i,j));
end;
end;
[R1,T1]=size(z1) ;
zz1=reshape(z1,R1*T1, 1);
% The transmitted data..................................
zz = [zz1];
            trel =poly2trellis(7,[171 133]); % Trellis
            data1=zz;
```

```
            zz = convenc(data1,trel);
s1=length(zz);
 para =128;
nd=6;        %number of information OFDM symbol for one
             loop
m1=2;        %Modulation level:QPSK
sr=250000;   %symbol rate
 br=sr.*ml;   %Bit rat per carrier
Ipoint = 8;  %Number of over samples
gilen=32;
flat=1;
fd=600;
epsilon=0.1;

    fftlen=para;
    noc=para;
    ofdm_length = para*nd*ml;   %Total no for one loop

%Dividing the image into blocks......................
nloops = ceil((length(zz))/ofdm_length );
new_data = nloops*ofdm_length ;
nzeros   = new_data   - length(zz);
input_data = [zz;zeros(nzeros,1)];
input_data2 = reshape(input_data ,ofdm_length ,nloops);

%Transmission ON FFT_OFDM............................

   for ebno=[0,2,4,6,8,10];

demodata1 = zeros(ofdm_length ,nloops);
for jj = 1: nloops          % loop for columns
    serdata = input_data2(:,jj)';
    demodata = fft_channel_offset_zg(serdata,para,nd,m
    1,gilen,fftlen,sr,ebno, br,epsilon);
   demodata1(:,jj) = demodata(:);   % the output of
   ofdm columns

end

%Received image......................................
 [Mr,Nr] = size(demodata1);
%  demodata2 = demodata1(:);
  yy      = reshape (demodata1,Mr*Nr,1); %
 part1    = yy(1:s1);
yy11=part1;
%%%%%%%%%%%%%%%%%%%%%%%%%%%%%%%%%%DECODING............
 yy11 = vitdec(yy11',trel,1,'term','hard'); % Decode.

yy1=reshape(yy11,[R1,T1]);
for i=1:M1
for j=1:N1
zn1(i,j)=num2str(yy1(i,j));
```

```
end;
end;
hn1=bin2dec(zn1);
gn1=col2im(hn1, [M,N], [M,N], 'distinct');
 %The Error between Trans.............................
   output_image = gn1/255;

output_image1=medfilt2( output_image);

%The Error between Trans...............................
 MSE1=sum(sum((double(f)/255-output_image).^2))/
 prod(size(f));
PSNR=10*log(1/MSE1)/log(10)
 MSE11=sum(sum((double(f)/255-output_image1).^2))/
 prod(size(f));
PSNR1=10*log(1/MSE11)/log(10);

   end

%Transmission ON DCT_OFDM.............................

   for ebno=[0,2,4,6,8,10];

demodata1 = zeros(ofdm_length ,nloops);
for jj = 1: nloops          % loop for columns
    serdata1 = input_data2(:,jj)';
    demodata = dct_channel_offset_zg(serdata1,para,nd,
    m1,gilen,fftlen,sr,ebno, br,epsilon);
   demodata1(:,jj) = demodata(:);   % the output of
   ofdm columns

end

%Received image.......................................
 [Mr,Nr] = size(demodata1);
%  demodata2 = demodata1(:);
  yy      = reshape (demodata1,Mr*Nr,1); %
 part1    = yy(1:s1);
yy11=part1;
%%%%%%%%%%%%%%%%%%%%%%%%%%%%%%%%DECODING...........
 yy11 = vitdec(yy11',trel,1,'term','hard'); % Decode.

yy1=reshape(yy11,[R1,T1]);
for i=1:M1
for j=1:N1
zn1(i,j)=num2str(yy1(i,j));
end;
end;
hn1=bin2dec(zn1);
gn1=col2im(hn1, [M,N], [M,N], 'distinct');
 %The Error between Trans.............................
   output_image = gn1/255;
```

```
output_image1=medfilt2( output_image);

%The Error between Trans.............................
 MSE1=sum(sum((double(f)/255-output_image).^2))/
 prod(size(f));
PSNR=10*log(1/MSE1)/log(10)
 MSE11=sum(sum((double(f)/255-output_image1).^2))/
 prod(size(f));
PSNR1=10*log(1/MSE11)/log(10);

   end

%Transmission ON DWT_OFDM............................

   for ebno=[0,2,4,6,8,10];

demodata1 = zeros(ofdm_length ,nloops);
for jj = 1: nloops           % loop for columns
    serdata1 = input_data2(:,jj)';
    demodata =dwt_channel_offset_zg(serdata1,para,nd,m
    1,gilen,fftlen,sr,ebno, br,epsilon);
   demodata1(:,jj) = demodata(:);   % the output of
   ofdm columns

end

%Received image......................................
 [Mr,Nr] = size(demodata1);
%  demodata2 = demodata1(:);
  yy      = reshape (demodata1,Mr*Nr,1); %
 part1   = yy(1:s1);
yy11=part1;
%%%%%%%%%%%%%%%%%%%%%%%%%%%%%%%%%%DECODING..........
 yy11 = vitdec(yy11',trel,1,'term','hard'); % Decode.

yy1=reshape(yy11,[R1,T1]);
for i=1:M1
for j=1:N1
zn1(i,j)=num2str(yy1(i,j));
end;
end;
hn1=bin2dec(zn1);
gn1=col2im(hn1, [M,N], [M,N], 'distinct');
 %The Error between Trans.............................
   output_image = gn1/255;

output_image1=medfilt2( output_image);

%The Error between Trans.............................
 MSE1=sum(sum((double(f)/255-output_image).^2))/
 prod(size(f));
PSNR=10*log(1/MSE1)/log(10)
```

```
 MSE11=sum(sum((double(f)/255-output_image1).^2))/
 prod(size(f));
PSNR1=10*log(1/MSE11)/log(10);

   end
```

%Image with no Encryption + OFDM + offset + cyclic prefix + over estimated channel........................
%Transmitter...................................
%Data Generation...............................

%Transmitter...................................

%Data Generation...............................

```
f=imread ('cameraman.tif' );
[M,N]=size(f) ;
g1 =im2col(f, [M,N], [M,N], 'distinct');
h1=dec2bin(double(g1));
[M1,N1]=size(h1) ;
z1=zeros (M1,N1) ;
for i=1:M1
for j=1:N1
z1(i,j)=str2num(h1(i,j));
end;
end;
[R1,T1]=size(z1) ;
zz1=reshape(z1,R1*T1, 1);
% The transmitted data................................
zz = [zz1];

%ENCODING........................................
            trel =poly2trellis(7,[171 133]); % Trellis
            data1=zz;
            zz = convenc(data1,trel);
s1=length(zz);
 para =128;
nd=6;       %number of information OFDM symbol for one
            loop
m1=2;       %Modulation level:QPSK
sr=250000;  %symbol rate
 br=sr.*m1;  %Bit rat per carrier
Ipoint = 8; %Number of over samples
gilen=32;
flat=1;
fd=600;

    fftlen=para;
    noc=para;
    ofdm_length = para*nd*m1;  N%Total no for one loop
```

```
%Dividing the image into blocks......................
nloops = ceil((length(zz))/ofdm_length );
new_data = nloops*ofdm_length ;
nzeros   = new_data  - length(zz);
input_data = [zz;zeros(nzeros,1)];
input_data2 = reshape(input_data ,ofdm_length ,nloops);

%Transmission ON FFT_OFDM.............................

   for ebno=[0,2,4,6,8,10];

demodata1 = zeros(ofdm_length ,nloops);
for jj = 1: nloops          % loop for columns
    serdata = input_data2(:,jj)';
    demodata = fft_channel_estimation_no_mapping_g(se
    rdata,para,nd,m1,gilen,fftlen,sr,ebno, br,fd,flat);
   demodata1(:,jj) = demodata(:);   % the output of
   ofdm columns

end

%Received image.......................................
 [Mr,Nr] = size(demodata1);
%  demodata2 = demodata1(:);
  yy      = reshape (demodata1,Mr*Nr,1); %
 part1   = yy(1:s1);
yy11=part1;
%DECODING.............................................
 yy11 = vitdec(yy11',trel,1,'term','hard'); % Decode.

yy1=reshape(yy11,[R1,T1]);
for i=1:M1
for j=1:N1
zn1(i,j)=num2str(yy1(i,j));
end;
end;
hn1=bin2dec(zn1);
gn1=col2im(hn1, [M,N], [M,N], 'distinct');
 %The Error between Trans.............................
   output_image = gn1/255;

output_image1=medfilt2( output_image);

%The Error between Trans..............................
 MSE1=sum(sum((double(f)/255-output_image).^2))/
 prod(size(f));
PSNR=10*log(1/MSE1)/log(10)
 MSE11=sum(sum((double(f)/255-output_image1).^2))/
 prod(size(f));
PSNR1=10*log(1/MSE11)/log(10);

   end
```

```
%Transmission ON DCT_OFDM.........................

for ebno=[0,2,4,6,8,10];

demodata1 = zeros(ofdm_length ,nloops);
for jj = 1: nloops          % loop for columns
    serdata = input_data2(:,jj)';
    demodata = dct_channel_estimation_no_mapping_g(se
    rdata,para,nd,m1,gilen,fftlen,sr,ebno, br,fd,flat);
   demodata1(:,jj) = demodata(:);    % the output of
   ofdm columns

end

%Received image..................................
 [Mr,Nr] = size(demodata1);
%  demodata2 = demodata1(:);
  yy       = reshape (demodata1,Mr*Nr,1); %
 part1    = yy(1:s1);
yy11=part1;
%DECODING........................................
 yy11 = vitdec(yy11',trel,1,'term','hard'); % Decode.

yy1=reshape(yy11,[R1,T1]);
for i=1:M1
for j=1:N1
zn1(i,j)=num2str(yy1(i,j));
end;
end;
hn1=bin2dec(zn1);
gn1=col2im(hn1, [M,N], [M,N], 'distinct');
 %The Error between Trans.........................
   output_image = gn1/255;

output_image1=medfilt2( output_image);

%The Error between Trans..........................
 MSE1=sum(sum((double(f)/255-output_image).^2))/
 prod(size(f));
PSNR=10*log(1/MSE1)/log(10)
 MSE11=sum(sum((double(f)/255-output_image1).^2))/
 prod(size(f));
PSNR1=10*log(1/MSE11)/log(10);

   end

   %Transmission ON DWT_OFDM.........................

   for ebno=[0,2,4,6,8,10];

demodata1 = zeros(ofdm_length ,nloops);
for jj = 1: nloops          % loop for columns
```

```
    serdata = input_data2(:,jj)';
    demodata = dwt_channel_estimation_no_mapping_g(se
    rdata,para,nd,m1,gilen,fftlen,sr,ebno, br,fd,flat);
   demodata1(:,jj) = demodata(:);   % the output of
   ofdm columns

end

%Received image.....................................
 [Mr,Nr] = size(demodata1);
%  demodata2 = demodata1(:);
  yy      = reshape (demodata1,Mr*Nr,1); %
 part1   = yy(1:s1);
yy11=part1;
%DECODING...........................................
 yy11 = vitdec(yy11',trel,1,'term','hard'); % Decode.

yy1=reshape(yy11,[R1,T1]);
for i=1:M1
for j=1:N1
zn1(i,j)=num2str(yy1(i,j));
end;
end;
hn1=bin2dec(zn1);
gn1=col2im(hn1, [M,N], [M,N], 'distinct');
 %The Error between Trans............................
   output_image = gn1/255;

output_image1=medfilt2( output_image);

%The Error between Trans.............................
 MSE1=sum(sum((double(f)/255-output_image).^2))/
 prod(size(f));
PSNR=10*log(1/MSE1)/log(10)
 MSE11=sum(sum((double(f)/255-output_image1).^2))/
 prod(size(f));
PSNR1=10*log(1/MSE11)/log(10);

   end

%Image with no Encryption + OFDM + offset + zero padding
+ over estimated channel............................
%Transmitter........................................
%Data Generation....................................

%Transmitter........................................

%Data Generation....................................

f=imread ('cameraman.tif' );
[M,N]=size(f) ;
g1 =im2col(f, [M,N], [M,N], 'distinct');
```

```
h1=dec2bin(double(g1));
[M1,N1]=size(h1) ;
z1=zeros (M1,N1) ;
for i=1:M1
for j=1:N1
z1(i,j)=str2num(h1(i,j));
end;
end;
[R1,T1]=size(z1) ;
zz1=reshape(z1,R1*T1, 1);
% The transmitted data.................................
zz = [zz1];

%ENCODING.............................................
            trel =poly2trellis(7,[171 133]); % Trellis
            data1=zz;
            zz = convenc(data1,trel);
s1=length(zz);
 para =128;
nd=6;       %number of information OFDM symbol for one
            loop
m1=2;       %Modulation level:QPSK
sr=250000;  %symbol rate
 br=sr.*m1;  %Bit rat per carrier
Ipoint = 8; %Number of over samples
gilen=32;
flat=1;
fd=600;

    fftlen=para;
    noc=para;
    ofdm_length = para*nd*m1;   %Total no for one loop

%Dividing the image into blocks.......................
nloops = ceil((length(zz))/ofdm_length );
new_data = nloops*ofdm_length ;
nzeros   = new_data  - length(zz);
input_data = [zz;zeros(nzeros,1)];
input_data2 = reshape(input_data ,ofdm_length ,nloops);

%Transmission ON FFT_OFDM.............................

   for ebno=[0,2,4,6,8,10];

demodata1 = zeros(ofdm_length ,nloops);
for jj = 1: nloops         % loop for columns
    serdata = input_data2(:,jj)';
    demodata = fft_channel_estimation_no_mapping_zg(se
    rdata,para,nd,m1,gilen,fftlen,sr,ebno, br,fd,flat);
  demodata1(:,jj) = demodata(:);   % the output of
  ofdm columns

end
```

```
%Received image.......................................
 [Mr,Nr] = size(demodata1);
%  demodata2 = demodata1(:);
  yy       = reshape (demodata1,Mr*Nr,1); %
 part1    = yy(1:s1);
yy11=part1;
%DECODING..............................................
 yy11 = vitdec(yy11',trel,1,'term','hard'); % Decode.

yy1=reshape(yy11,[R1,T1]);
for i=1:M1
for j=1:N1
zn1(i,j)=num2str(yy1(i,j));
end;
end;
hn1=bin2dec(zn1);
gn1=col2im(hn1, [M,N], [M,N], 'distinct');
 %The Error between Trans..............................
   output_image = gn1/255;

output_image1=medfilt2( output_image);

%The Error between Trans...............................
 MSE1=sum(sum((double(f)/255-output_image).^2))/
 prod(size(f));
PSNR=10*log(1/MSE1)/log(10)
 MSE11=sum(sum((double(f)/255-output_image1).^2))/
 prod(size(f));
PSNR1=10*log(1/MSE11)/log(10);

   end

   %Transmission ON DCT_OFDM...........................

   for ebno=[0,2,4,6,8,10];

demodata1 = zeros(ofdm_length ,nloops);
for jj = 1: nloops          % loop for columns
    serdata = input_data2(:,jj)';
    demodata = dct_channel_estimation_no_mapping_zg(se
    rdata,para,nd,m1,gilen,fftlen,sr,ebno, br,fd,flat);
   demodata1(:,jj) = demodata(:);   % the output of
   ofdm columns

end

%Received image.......................................
 [Mr,Nr] = size(demodata1);
%  demodata2 = demodata1(:);
  yy       = reshape (demodata1,Mr*Nr,1); %
 part1    = yy(1:s1);
yy11=part1;
```

```
%DECODING......................................
 yy11 = vitdec(yy11',trel,1,'term','hard'); % Decode.

yy1=reshape(yy11,[R1,T1]);
for i=1:M1
for j=1:N1
zn1(i,j)=num2str(yy1(i,j));
end;
end;
hn1=bin2dec(zn1);
gn1=col2im(hn1, [M,N], [M,N], 'distinct');
 %The Error between Trans.........................
   output_image = gn1/255;

output_image1=medfilt2( output_image);

%The Error between Trans..........................
 MSE1=sum(sum((double(f)/255-output_image).^2))/
 prod(size(f));
PSNR=10*log(1/MSE1)/log(10)
 MSE11=sum(sum((double(f)/255-output_image1).^2))/
 prod(size(f));
PSNR1=10*log(1/MSE11)/log(10);

   end

   %Transmission ON DWT_OFDM.......................

   for ebno=[0,2,4,6,8,10];

demodata1 = zeros(ofdm_length ,nloops);
for jj = 1: nloops          % loop for columns
    serdata = input_data2(:,jj)';
    demodata = dwt_channel_estimation_no_mapping_zg(se
    rdata,para,nd,m1,gilen,fftlen,sr,ebno, br,fd,flat);
   demodata1(:,jj) = demodata(:);    % the output of
   ofdm columns

end

%Received image................................
 [Mr,Nr] = size(demodata1);
%  demodata2 = demodata1(:);
  yy      = reshape (demodata1,Mr*Nr,1); %
 part1   = yy(1:s1);
yy11=part1;
%DECODING......................................
 yy11 = vitdec(yy11',trel,1,'term','hard'); % Decode.

yy1=reshape(yy11,[R1,T1]);
for i=1:M1
for j=1:N1
```

```
zn1(i,j)=num2str(yy1(i,j));
end;
end;
hn1=bin2dec(zn1);
gn1=col2im(hn1, [M,N], [M,N], 'distinct');
 %The Error between Trans..........................
   output_image = gn1/255;

output_image1=medfilt2( output_image);

%The Error between Trans...........................
 MSE1=sum(sum((double(f)/255-output_image).^2))/
 prod(size(f));
PSNR=10*log(1/MSE1)/log(10)
 MSE11=sum(sum((double(f)/255-output_image1).^2))/
 prod(size(f));
PSNR1=10*log(1/MSE11)/log(10);

   end


%%%%%%%%%%%%%%%%%%%%%%%%%%%%%%%%%%%%%%%%%%%%%%%%%%%%
function outdemodata = fft_channel(serdata,para,nd,m1,
gilen,fftlen,sr,ebno, br)

%serial to parallel convertion.....................

paradata = reshape(serdata,para,nd*m1);
%QPSK modulation...................................
[ich,qch] = qpskmod(paradata,para,nd,m1);
% [ich0,qch0] = compoversamp(ich01,qch01,length(ich01)
,Ipoint);
kmod = 1/sqrt(2);
ich1 = ich.*kmod;
qch1 = qch.*kmod;
%IFFT..............................................
x = ich1 + qch1.*j;
y = ifft(x);
ich2 = real (y);
qch2 = imag (y);

%Gaurd interval insertion..........................

[ich3,qch3] = giins(ich2,qch2,fftlen,gilen,nd);
fftlen2 = fftlen + gilen;

%Attenuation Calculation...........................

spow = sum(ich3.^2+qch3.^2)/nd./para;
attn = 0.5*spow*sr/br*10.^(-ebno/10);
attn = sqrt (attn);
```

```
%Receiever.........................................

%AWGN addition......................................

 [ich4,qch4] = comb(ich3,qch3,attn);

%Guard interval removal.............................
 [ich5,qch5] = girem (ich4,qch4,fftlen2,gilen,nd);
%FFT...............................................
rx = ich5 + qch5.*j;
ry = fft(rx);
ich6 = real (ry);
qch6 = imag (ry);

%Demodulation.......................................
 ich7 = ich6./ kmod;
 qch7 = qch6./ kmod;
 outdemodata  = qpskdemod (ich7,qch7,para,nd,m1);


function outdemodata = dct_channel(serdata,para,nd,m1,
gilen,dctlen,sr,ebno, br)

%serial to parallel convertion......................

paradata = reshape(serdata,para,nd*m1);
%QPSK modulation....................................
[ich,qch] = qpskmod(paradata,para,nd,m1);
% [ich0,qch0] = compoversamp(ich01,qch01,length(ich01)
,Ipoint);
kmod = 1/sqrt(2);
ich1 = ich.*kmod;
qch1 = qch.*kmod;
%IDCT...............................................
x = ich1 + qch1.*j;
y = idct(x);
ich2 = real (y);
qch2 = imag (y);

%Gaurd interval insertion...........................

[ich3,qch3] = giins(ich2,qch2,dctlen,gilen,nd);
dctlen2 = dctlen + gilen;

%Attenuation Calculation............................

spow = sum(ich3.^2+qch3.^2)/nd./para;
attn = 0.5*spow*sr/br*10.^(-ebno/10);
attn = sqrt (attn);

%Receiever..........................................
```

```
%AWGN addition.........................................

 [ich4,qch4] = comb(ich3,qch3,attn);

%Guard interval removal................................
 [ich5,qch5] = girem (ich4,qch4,dctlen2,gilen,nd);
%DCT...................................................
rx = ich5 + qch5.*j;
ry = dct(rx);
ich6 = real (ry);
qch6 = imag (ry);

%Demodulation..........................................
 ich7 = ich6./ kmod;
 qch7 = qch6./ kmod;
 outdemodata  = qpskdemod (ich7,qch7,para,nd,m1);


function outdemodata = dwt_channel(serdata,para,nd,m1,
gilen,dctlen,sr,ebno, br)

%serial to parallel convertion.........................

paradata = reshape(serdata,para,nd*m1);
%QPSK modulation.......................................
[ich,qch] = qpskmod(paradata,para,nd,m1);
% [ich0,qch0] = compoversamp(ich01,qch01,length(ich01)
,Ipoint);
kmod = 1/sqrt(2);
ich1 = ich.*kmod;
qch1 = qch.*kmod;
%IDWT..................................................
x = ich1 + qch1.*j;
y = wavelet('D6',-1,x,'zpd');    % Invert 5 stages
%;    % 2D wavelet transform
%   R = wavelet('2D CDF 9/7',-2,Y); % Recover X from Y
% Forward transform with 5 stages

ich2 = real (y);
qch2 = imag (y);

%Gaurd interval insertion..............................

[ich3,qch3] = giins(ich2,qch2,dctlen,gilen,nd);
dctlen2 = dctlen + gilen;

%Attenuation Calculation...............................

spow = sum(ich3.^2+qch3.^2)/nd./para;
attn = 0.5*spow*sr/br*10.^(-ebno/10);
attn = sqrt (attn);
```

```
%Receiever.........................................

%AWGN addition.....................................

 [ich4,qch4] = comb(ich3,qch3,attn);

%Guard interval removal............................
 [ich5,qch5] = girem (ich4,qch4,dctlen2,gilen,nd);
%DWT...............................................
rx = ich5 + qch5.*j;
ry =wavelet('D6',1,rx,'zpd');;  % 2D wavelet transform
%  R = wavelet('2D CDF 9/7',-2,Y);  % Recover X from Y;
ich6 = real (ry);
qch6 = imag (ry);

%Demodulation......................................
 ich7 = ich6./ kmod;
 qch7 = qch6./ kmod;
 outdemodata  = qpskdemod (ich7,qch7,para,nd,m1);


function outdemodata = fft_channel_clipping_companding
(serdata,para,nd,m1,gilen,fftlen,sr,ebno, br)

paradata = reshape(serdata,para,nd*m1);

%QPSK modulation...................................
[ich,qch] = qpskmod(paradata,para,nd,m1);
% [ich0,qch0] = compoversamp(ich01,qch01,length(ich01)
,Ipoint);
kmod = 1/sqrt(2);
ich1 = ich.*kmod;
qch1 = qch.*kmod;

%IFFT..............................................
x = ich1 + qch1.*i;
y = ifft(x);
CR=4;
      clipping_threshold=(10^(CR/10))*sqrt(mean(abs
      (y).^2));
       tx_signal_Ang = angle(y);

       for ii=1:length(y)
        if y(ii)== y(ii);
            y(ii)=y(ii);

        elseif   abs(y(ii)).^2> clipping_threshold
            y(ii)=clipping_threshold.*exp(sqrt(-1)*tx_
            signal_Ang(ii));

        end
       end
```

```
%companding...............................
u=4;

         tx_lfdma_max=max(abs(y(1:nd*para)));
         tx_lfdma_Abs = abs(y(1:nd*para));  % tx data
         amplitude

         TxSamples_lfdma1= tx_lfdma_max*((log10(1+u*
(tx_lfdma_Abs./tx_lfdma_max)))/log10(u+1)).*sign(y(1:
nd*para));

         tx_ifdma_max=max(abs(y(1:nd*para)));
         tx_ifdma_Abs = abs(y(1:nd*para));  % tx data
         amplitude
         yy= tx_ifdma_max*((log10(1+u*(tx_ifdma_Abs./
tx_ifdma_max)))/log10(u+1)).*sign(y(1:nd*para));

ich3 = real (yy);
qch3 = imag (yy);

%Gaurd interval insertion.............................

[ich4,qch4] = giins(ich3,qch3,fftlen,gilen,nd);
fftlen2 = fftlen + gilen;

%Attenuation Calculation..............................

spow = sum(ich4.^2+qch4.^2)/nd./para;
attn = 0.5*spow*sr/br*10.^(-ebno/10);
attn = sqrt (attn);
%fading...............................................
%%******************** Create Rayleigh fading channel
object. ******************

%***************** AWGN addition *****************
%Receiever............................................

%AWGN addition........................................

 [ich5,qch5] = comb(ich4,qch4,attn);
 %perfect fading compensation.........................

%Guard interval removal...............................
 [ich6,qch6] = girem (ich5,qch5,fftlen2,gilen,nd);
%FFT..................................................
rx = ich6 + qch6.*j;

         %%%%%%%%%%% Expanding
         rx_lfdma_Abs = abs( rx);
        r_lfdma_max=tx_lfdma_max;
```

```
RxSamples_lfdma=(r_lfdma_max/u)*(exp(log10(1+u)*2.3025
85093*rx_lfdma_Abs./r_lfdma_max)-1).*sign(rx);

        rx_ifdma_Abs = abs(rx);
        r_ifdma_max=tx_ifdma_max;

rxx=(r_ifdma_max/u)*(exp(log10(1+u)*2.302585093*rx_
ifdma_Abs./r_ifdma_max)-1).*sign(rx);

ry = fft(rxx);
ich7 = real (ry);
qch7 = imag (ry);

%Demodulation.......................................
 ich10 = ich7./ kmod;
 qch10 = qch7./ kmod;
 outdemodata  = qpskdemod (ich10,qch10,para,nd,m1);


function outdemodata = dct_channel_clipping_companding
(serdata,para,nd,m1,gilen,dctlen,sr,ebno, br)

%serial to parallel convertion......................

paradata = reshape(serdata,para,nd*m1);
%QPSK modulation....................................
[ich,qch] = qpskmod(paradata,para,nd,m1);
% [ich0,qch0] = compoversamp(ich01,qch01,length(ich01)
,Ipoint);
kmod = 1/sqrt(2);
ich1 = ich.*kmod;
qch1 = qch.*kmod;
%IDCT...............................................
x = ich1 + qch1.*j;
y = idct(x);

CR=4;
      clipping_threshold=(10^(CR/10))*sqrt(mean(abs
      (y).^2));
       tx_signal_Ang = angle(y);

       for ii=1:length(y)
        if y(ii)== y(ii);
            y(ii)=y(ii);

        elseif   abs(y(ii)).^2> clipping_threshold
            y(ii)=clipping_threshold.*exp(sqrt(-1)*tx_
            signal_Ang(ii));

        end
       end
```

```
%companding................................
u=4;
         tx_lfdma_max=max(abs(y(1:nd*para)));
         tx_lfdma_Abs = abs(y(1:nd*para));  % tx data
         amplitude

         TxSamples_lfdma1= tx_lfdma_max*((log10(1+u*
(tx_lfdma_Abs./tx_lfdma_max)))/log10(u+1)).*sign(y(1:
nd*para));

         tx_ifdma_max=max(abs(y(1:nd*para)));
         tx_ifdma_Abs = abs(y(1:nd*para));  % tx data
         amplitude
         yy= tx_ifdma_max*((log10(1+u*(tx_ifdma_Abs./
tx_ifdma_max)))/log10(u+1)).*sign(y(1:nd*para));

ich2 = real (yy);
qch2 = imag (yy);

%Gaurd interval insertion............................

[ich3,qch3] = giins(ich2,qch2,dctlen,gilen,nd);
dctlen2 = dctlen + gilen;

%Attenuation Calculation.............................

spow = sum(ich3.^2+qch3.^2)/nd./para;
attn = 0.5*spow*sr/br*10.^(-ebno/10);
attn = sqrt (attn);

%Receiever...........................................

%AWGN addition.......................................

 [ich4,qch4] = comb(ich3,qch3,attn);

%Guard interval removal..............................
 [ich5,qch5] = girem (ich4,qch4,dctlen2,gilen,nd);
%DCT.................................................
rx = ich5 + qch5.*j;

         %%%%%%%%%%% Expanding
         rx_lfdma_Abs = abs( rx);
        r_lfdma_max=tx_lfdma_max;

RxSamples_lfdma=(r_lfdma_max/u)*(exp(log10(1+u)*2.3025
85093*rx_lfdma_Abs./r_lfdma_max)-1).*sign(rx);

        rx_ifdma_Abs = abs(rx);
        r_ifdma_max=tx_ifdma_max;
```

```
rxx=(r_ifdma_max/u)*(exp(log10(1+u)*2.302585093*rx_
ifdma_Abs./r_ifdma_max)-1).*sign(rx);

ry = dct(rxx);
ich6 = real (ry);
qch6 = imag (ry);

%Demodulation...................................
 ich7 = ich6./ kmod;
 qch7 = qch6./ kmod;
 outdemodata  = qpskdemod (ich7,qch7,para,nd,m1);


function outdemodata = dwt_channel_clipping_companding
(serdata,para,nd,m1,gilen,dctlen,sr,ebno, br)

%serial to parallel convertion.......................

paradata = reshape(serdata,para,nd*m1);
%QPSK modulation...................................
[ich,qch] = qpskmod(paradata,para,nd,m1);
% [ich0,qch0] = compoversamp(ich01,qch01,length(ich01)
,Ipoint);
kmod = 1/sqrt(2);
ich1 = ich.*kmod;
qch1 = qch.*kmod;
%IDWT.....................................
x = ich1 + qch1.*j;
y = wavelet('D6',-1,x,'zpd');   % Invert 5 stages
%;    % 2D wavelet transform
%   R = wavelet('2D CDF 9/7',-2,Y);   % Recover X from
Y % Forward transform with 5 stages
CR=4;
      clipping_threshold=(10^(CR/10))*sqrt(mean(abs
      (y).^2));
       tx_signal_Ang = angle(y);

       for ii=1:length(y)
        if y(ii)== y(ii);
            y(ii)=y(ii);


        elseif   abs(y(ii)).^2> clipping_threshold
            y(ii)=clipping_threshold.*exp(sqrt(-1)*tx_
            signal_Ang(ii));

        end
       end
%companding............................
u=4;
        tx_lfdma_max=max(abs(y(1:nd*para)));
```

```
         tx_lfdma_Abs = abs(y(1:nd*para));  % tx data
         amplitude

         TxSamples_lfdma1= tx_lfdma_max*((log10(1+u*
(tx_lfdma_Abs./tx_lfdma_max)))/log10(u+1)).*sign(y(1:
nd*para));

         tx_ifdma_max=max(abs(y(1:nd*para)));
         tx_ifdma_Abs = abs(y(1:nd*para));  % tx data
         amplitude
         yy= tx_ifdma_max*((log10(1+u*(tx_ifdma_Abs./
tx_ifdma_max)))/log10(u+1)).*sign(y(1:nd*para));

ich2 = real (yy);
qch2 = imag (yy);

%Gaurd interval insertion..............................

[ich3,qch3] = giins(ich2,qch2,dctlen,gilen,nd);
dctlen2 = dctlen + gilen;

%Attenuation Calculation...............................

spow = sum(ich3.^2+qch3.^2)/nd./para;
attn = 0.5*spow*sr/br*10.^(-ebno/10);
attn = sqrt (attn);

%Receiever.............................................

%AWGN addition.........................................

 [ich4,qch4] = comb(ich3,qch3,attn);

%Guard interval removal................................
 [ich5,qch5] = girem (ich4,qch4,dctlen2,gilen,nd);
%DWT...................................................
rx = ich5 + qch5.*j;

         %%%%%%%%%%% Expanding
         rx_lfdma_Abs = abs( rx);
        r_lfdma_max=tx_lfdma_max;

RxSamples_lfdma=(r_lfdma_max/u)*(exp(log10(1+u)*2.3025
85093*rx_lfdma_Abs./r_lfdma_max)-1).*sign(rx);

        rx_ifdma_Abs = abs(rx);
        r_ifdma_max=tx_ifdma_max;

rxx=(r_ifdma_max/u)*(exp(log10(1+u)*2.302585093*rx_
ifdma_Abs./r_ifdma_max)-1).*sign(rx);

ry =wavelet('D6',1,rxx,'zpd');; % 2D wavelet transform
```

```
%   R = wavelet('2D CDF 9/7',-2,Y); % Recover X from Y;
ich6 = real (ry);
qch6 = imag (ry);

%Demodulation.....................................
 ich7 = ich6./ kmod;
 qch7 = qch6./ kmod;
 outdemodata  = qpskdemod (ich7,qch7,para,nd,m1);


function outdemodata = fft_channel_estimation_no_
mapping_clipping(serdata,para,nd,m1,gilen,fftlen,sr,e
bno, br,fd,flat)

paradata = reshape(serdata,para,nd*m1);
%QPSK modulation..................................
[ich,qch] = qpskmod(paradata,para,nd,m1);
% [ich0,qch0] = compoversamp(ich01,qch01,length(ich01)
,Ipoint);
kmod = 1/sqrt(2);
ich1 = ich.*kmod;
qch1 = qch.*kmod;
%channel estimation data generation..................
kndata=zeros(1,fftlen);
kndata0=2.*(rand(1,para)<0.5)-1;
kndata(1:para/2)=kndata0(1:para/2);
kndata((para/2)+1:para)=kndata0((para/2)+1:para);
ceich=kndata;
ceqch=zeros(1,para);
%data mapping.....................................
ich2=[ceich.' ich1];
qch2=[ceqch.' qch1];

%IFFT.............................................
x = ich2 + qch2.*i;

y = ifft(x);

CR=4;
      clipping_threshold=(10^(CR/10))*sqrt(mean(abs
      (y).^2));
       tx_signal_Ang = angle(y);

       for ii=1:length(y)
        if y(ii)== y(ii);
            y(ii)=y(ii);

        elseif   abs(y(ii)).^2> clipping_threshold
            y(ii)=clipping_threshold.*exp(sqrt(-1)*tx_
            signal_Ang(ii));

        end
       end
```

```
ich3 = real (y);
qch3 = imag (y);

%Gaurd interval insertion...........................

[ich4,qch4] = giins(ich3,qch3,fftlen,gilen,nd+1);
fftlen2 = fftlen + gilen;

%Attenuation Calculation............................

spow = sum(ich4.^2+qch4.^2)/nd./para;
attn = 0.5*spow*sr/br*10.^(-ebno/10);
attn = sqrt (attn);
%fading.............................................
%******************** Create Rayleigh fading channel
object. ******************
tstp=1/sr/(fftlen+gilen);
itau=[0,2,3,4];
dlvl1=[0,10,20,25];
n0=[6,7,6,7];
th1=[0,0,0,0];
itnd1=[1000,2000,3000,4000];
now1=4;
itnd0=nd*(fftlen+gilen)*20;
[ifade,qfade,ramp,rcos,rsin]=sefade(ich4,qch4,itau,dlv
l1,th1,n0,itnd1,now1,length(ich4),tstp,fd,flat);
itnd1=itnd1+itnd0;
ich4=ifade;
qch4=qfade;

%***************** AWGN addition ********************
%Receiever..........................................

%AWGN addition......................................

 [ich5,qch5] = comb(ich4,qch4,attn);
 %perfect fading compensation.......................
 ifade2=1./ramp.*(rcos(1,:).*ich5+rsin(1,:).*qch5);
 qfade2=1./ramp.*(-rsin(1,:).*ich5+rcos(1,:).*qch5);
 ich5=ifade2;
 qch5=qfade2;

%Guard interval removal.............................
 [ich6,qch6] = girem (ich5,qch5,fftlen2,gilen,nd+1);
%FFT................................................
rx = ich6 + qch6.*j;
ry = fft(rx);
ich7 = real (ry);
qch7 = imag (ry);
%fading compensation by channel estimation symbol.....
ce=1;
```

```
ice0=ich2(:,ce);
qce0=qch2(:,ce);
ice1=ich7(:,ce);
qce1=qch7(:,ce);
%calculate reverse rotation...........................
iv=real((1./(ice1.^2+qce1.^2)).*(ice0+i.*qce0).*(ice1-
i.*qce1));
qv=imag((1./(ice1.^2+qce1.^2)).*(ice0+i.*qce0).*(ice1-
i.*qce1));
%matrix for reverse rotation..........................
ieqv1=[iv iv iv iv iv iv iv];
qeqv1=[qv qv qv qv qv qv qv];
%reverse rptation.....................................
icompen=real((ich7+i.*qch7).*(ieqv1+i.*qeqv1));
qcompen=imag((ich7+i.*qch7).*(ieqv1+i.*qeqv1));
ich7=icompen;
qch7=qcompen;
%channel estimation symbol removal...................
knd=1;               %number of known channel
                     estimation ofdm symbol
ich9=ich7(:,knd+1:nd+1);
qch9=qch7(:,knd+1:nd+1);

%Demodulation.........................................

 ich10 = ich9./ kmod;
 qch10 = qch9./ kmod;
 outdemodata  = qpskdemod (ich10,qch10,para,nd,m1);


function outdemodata = dct_channel_estimation_no_
mapping_clipping(serdata,para,nd,m1,gilen,fftlen,sr,e
bno, br,fd,flat)

paradata = reshape(serdata,para,nd*m1);

%QPSK modulation......................................
[ich,qch] = qpskmod(paradata,para,nd,m1);
% [ich0,qch0] = compoversamp(ich01,qch01,length(ich01)
,Ipoint);
kmod = 1/sqrt(2);
ich1 = ich.*kmod;
qch1 = qch.*kmod;
%channel estimation data generation..................
kndata=zeros(1,fftlen);
kndata0=2.*(rand(1,para)<0.5)-1;
kndata(1:para/2)=kndata0(1:para/2);
kndata((para/2)+1:para)=kndata0((para/2)+1:para);
ceich=kndata;
ceqch=zeros(1,para);
%data mapping.........................................
```

```
ich2=[ceich.' ich1];
qch2=[ceqch.' qch1];

%IDCT.......................................................
x = ich2 + qch2.*i;
y = idct(x);
CR=4;
      clipping_threshold=(10^(CR/10))*sqrt(mean(abs
      (y).^2));
       tx_signal_Ang = angle(y);

      for ii=1:length(y)
        if y(ii)== y(ii);
            y(ii)=y(ii);

        elseif   abs(y(ii)).^2> clipping_threshold
            y(ii)=clipping_threshold.*exp(sqrt(-1)*tx_
            signal_Ang(ii));

        end
      end

ich3 = real (y);
qch3 = imag (y);

%Gaurd interval insertion...............................

[ich4,qch4] = giins(ich3,qch3,fftlen,gilen,nd+1);
fftlen2 = fftlen + gilen;

%Attenuation Calculation................................

spow = sum(ich4.^2+qch4.^2)/nd./para;
attn = 0.5*spow*sr/br*10.^(-ebno/10);
attn = sqrt (attn);
%fading.....................................................
%******************** Create Rayleigh fading channel
object. *******************
tstp=1/sr/(fftlen+gilen);
itau=[0,2,3,4];
dlvl1=[0,10,20,25];
n0=[6,7,6,7];
th1=[0,0,0,0];
itnd1=[1000,2000,3000,4000];
now1=4;
itnd0=nd*(fftlen+gilen)*20;
[ifade,qfade,ramp,rcos,rsin]=sefade(ich4,qch4,itau,dlv
l1,th1,n0,itnd1,now1,length(ich4),tstp,fd,flat);
itnd1=itnd1+itnd0;
ich4=ifade;
qch4=qfade;
```

```
%***************** AWGN addition ********************
%Receiever.......................................

%AWGN addition...................................

 [ich5,qch5] = comb(ich4,qch4,attn);
 %perfect fading compensation.........................
 ifade2=1./ramp.*(rcos(1,:).*ich5+rsin(1,:).*qch5);
 qfade2=1./ramp.*(-rsin(1,:).*ich5+rcos(1,:).*qch5);
 ich5=ifade2;
 qch5=qfade2;

%Guard interval removal..........................

 [ich6,qch6] = girem (ich5,qch5,fftlen2,gilen,nd+1);
%DCT..............................................
rx = ich6 + qch6.*j;
ry = dct(rx);
ich7 = real (ry);
qch7 = imag (ry);
%fading compensation by channel estimation symbol.....
ce=1;
ice0=ich2(:,ce);
qce0=qch2(:,ce);
ice1=ich7(:,ce);
qce1=qch7(:,ce);
%calculate reverse rotation.........................
iv=real((1./(ice1.^2+qce1.^2)).*(ice0+i.*qce0).*(ice1-
i.*qce1));
qv=imag((1./(ice1.^2+qce1.^2)).*(ice0+i.*qce0).*(ice1-
i.*qce1));
%matrix for reverse rotation........................
ieqv1=[iv iv iv iv iv iv iv];
qeqv1=[qv qv qv qv qv qv qv];
%reverse rptation......................
icompen=real((ich7+i.*qch7).*(ieqv1+i.*qeqv1));
qcompen=imag((ich7+i.*qch7).*(ieqv1+i.*qeqv1));
ich7=icompen;
qch7=qcompen;
%channel estimation symbol removal..................
knd=1;                  %number of known channel
                        estimation ofdm symbol
ich9=ich7(:,knd+1:nd+1);
qch9=qch7(:,knd+1:nd+1);

%Demodulation......................................
 ich10 = ich9./ kmod;
 qch10 = qch9./ kmod;
 outdemodata  = qpskdemod (ich10,qch10,para,nd,m1);
```

```
function outdemodata = dwt_channel_estimation_no_
mapping_clipping(serdata,para,nd,m1,gilen,fftlen,sr,e
bno, br,fd,flat)

paradata = reshape(serdata,para,nd*m1);
%QPSK modulation.................................
[ich,qch] = qpskmod(paradata,para,nd,m1);
% [ich0,qch0] = compoversamp(ich01,qch01,length(ich01)
,Ipoint);
kmod = 1/sqrt(2);
ich1 = ich.*kmod;
qch1 = qch.*kmod;
%channel estimation data generation..................
kndata=zeros(1,fftlen);
kndata0=2.*(rand(1,para)<0.5)-1;
kndata(1:para/2)=kndata0(1:para/2);
kndata((para/2)+1:para)=kndata0((para/2)+1:para);
ceich=kndata;
ceqch=zeros(1,para);
%data mapping...................................
ich2=[ceich.' ich1];
qch2=[ceqch.' qch1];

%IDWT...........................................
x = ich2 + qch2.*j;
y = wavelet('D6',-1,x,'zpd');    % Invert 5 stages
%;    % 2D wavelet transform
%   R = wavelet('2D CDF 9/7',-2,Y);   % Recover X from
Y % Forward transform with 5 stages
CR=4;
      clipping_threshold=(10^(CR/10))*sqrt(mean(abs
      (y).^2));
       tx_signal_Ang = angle(y);

       for ii=1:length(y)
        if y(ii)== y(ii);
            y(ii)=y(ii);

        elseif   abs(y(ii)).^2> clipping_threshold
            y(ii)=clipping_threshold.*exp(sqrt(-1)*tx_
            signal_Ang(ii));
        end
       end

ich3 = real (y);
qch3 = imag (y);

%Gaurd interval insertion.........................

[ich4,qch4] = giins(ich3,qch3,fftlen,gilen,nd+1);
fftlen2 = fftlen + gilen;
```

```
%Attenuation Calculation..............................

spow = sum(ich4.^2+qch4.^2)/nd./para;
attn = 0.5*spow*sr/br*10.^(-ebno/10);
attn = sqrt (attn);
%fading...............................................
%******************** Create Rayleigh fading channel
object. ********************
tstp=1/sr/(fftlen+gilen);
itau=[0,2,3,4];
dlvl1=[0,10,20,25];
n0=[6,7,6,7];
th1=[0,0,0,0];
itnd1=[1000,2000,3000,4000];
now1=4;
itnd0=nd*(fftlen+gilen)*20;
[ifade,qfade,ramp,rcos,rsin]=sefade(ich4,qch4,itau,dlv
l1,th1,n0,itnd1,now1,length(ich4),tstp,fd,flat);
itnd1=itnd1+itnd0;
ich4=ifade;
qch4=qfade;

%***************** AWGN addition *******************
%Receiever............................................

%AWGN addition........................................

 [ich5,qch5] = comb(ich4,qch4,attn);
 %perfect fading compensation.........................
 ifade2=1./ramp.*(rcos(1,:).*ich5+rsin(1,:).*qch5);
 qfade2=1./ramp.*(-rsin(1,:).*ich5+rcos(1,:).*qch5);
 ich5=ifade2;
 qch5=qfade2;

%Guard interval removal...............................
 [ich6,qch6] = girem (ich5,qch5,fftlen2,gilen,nd+1);
%DWT..................................................
rx = ich6 + qch6.*j;
ry =wavelet('D6',1,rx,'zpd');; % 2D wavelet transform
%   R = wavelet('2D CDF 9/7',-2,Y); % Recover X from Y;
ich7 = real (ry);
qch7 = imag (ry);

%fading compensation by channel estimation symbol.....
ce=1;
ice0=ich2(:,ce);
qce0=qch2(:,ce);
ice1=ich7(:,ce);
qce1=qch7(:,ce);
%calculate reverse rotation...........................
iv=real((1./(ice1.^2+qce1.^2)).*(ice0+i.*qce0).*(ice1-
```

```
i.*qce1));
qv=imag((1./(ice1.^2+qce1.^2)).*(ice0+i.*qce0).*(ice1-
i.*qce1));
%matrix for reverse rotation.........................
ieqv1=[iv iv iv iv iv iv iv];
qeqv1=[qv qv qv qv qv qv qv];
%reverse rotation......................
icompen=real((ich7+i.*qch7).*(ieqv1+i.*qeqv1));
qcompen=imag((ich7+i.*qch7).*(ieqv1+i.*qeqv1));
ich7=icompen;
qch7=qcompen;
%channel estimation symbol removal....................
knd=1;                  %number of known channel
                        estimation ofdm symbol
ich9=ich7(:,knd+1:nd+1);
qch9=qch7(:,knd+1:nd+1);

%Demodulation...................................
 ich10 = ich9./ kmod;
 qch10 = qch9./ kmod;
 outdemodata  = qpskdemod (ich10,qch10,para,nd,m1);


function outdemodata = fft_channel_offset_g(serdata,pa
ra,nd,m1,gilen,fftlen,sr,ebno, br,epsilon)

%serial to parallel convertion.......................

paradata = reshape(serdata,para,nd*m1);

%QPSK modulation...................................
[ich,qch] = qpskmod(paradata,para,nd,m1);

% [ich0,qch0] = compoversamp(ich01,qch01,length(ich01)
,Ipoint);
kmod = 1/sqrt(2);
ich1 = ich.*kmod;
qch1 = qch.*kmod;

%IFFT.............................................
x = ich1 + qch1.*j;
y = ifft(x);
ich2 = real (y);
qch2 = imag (y);

%Gaurd interval insertion..........................
[ich3,qch3] = giins1(ich2,qch2,fftlen,gilen,nd);
fftlen2 = fftlen + gilen;

%Attenuation Calculation...........................

spow = sum(ich3.^2+qch3.^2)/nd./para;
```

```
attn = 0.5*spow*sr/br*10.^(-ebno/10);
attn = sqrt (attn);

%offset.........................................
n=para/2;
offset=exp(j*2*pi*n*epsilon/para);
i_rx_signal = ich3.*offset;
q_rx_signal = qch3.*offset;
ich3a = i_rx_signal ;
qch3a = q_rx_signal ;

%Receiever......................................
%AWGN addition..................................
 [ich4,qch4] = comb(ich3a,qch3a,attn);

%Guard interval removal.........................
 [ich5,qch5] = girem1 (ich4,qch4,fftlen2,gilen,nd);

%FFT............................................
rx = ich5 + qch5.*j;
ry = fft(rx);
ich6 = real (ry);
qch6 = imag (ry);

%Demodulation...................................
 ich7 = ich6./ kmod;
 qch7 = qch6./ kmod;
 outdemodata  = qpskdemod (ich7,qch7,para,nd,m1);


function outdemodata = dct_channel_offset_g(serdata,pa
ra,nd,m1,gilen,fftlen,sr,ebno, br,epsilon)

%serial to parallel convertion..................
paradata = reshape(serdata,para,nd*m1);

%QPSK modulation................................
[ich,qch] = qpskmod(paradata,para,nd,m1);

% [ich0,qch0] = compoversamp(ich01,qch01,length(ich01)
,Ipoint);
kmod = 1/sqrt(2);
ich1 = ich.*kmod;
qch1 = qch.*kmod;

%IDCT...........................................
x = ich1 + qch1.*j;
y = idct(x);
ich2 = real (y);
qch2 = imag (y);

%Gaurd interval insertion.......................
[ich3,qch3] = giins1(ich2,qch2,fftlen,gilen,nd);
fftlen2 = fftlen + gilen;
```

```
%Attenuation Calculation...............................
spow = sum(ich3.^2+qch3.^2)/nd./para;
attn = 0.5*spow*sr/br*10.^(-ebno/10);
attn = sqrt (attn);

%offset.................................................
n=para/2;
offset=exp(j*2*pi*n*epsilon/para);
i_rx_signal = ich3.*offset;
q_rx_signal = qch3.*offset;
ich3a = i_rx_signal ;
qch3a = q_rx_signal ;

%Receiever..............................................
%AWGN addition..........................................
 [ich4,qch4] = comb(ich3a,qch3a,attn);

%Guard interval removal.................................
 [ich5,qch5] = girem1 (ich4,qch4,fftlen2,gilen,nd);

%DCT....................................................
rx = ich5 + qch5.*j;
ry = dct(rx);
ich6 = real (ry);
qch6 = imag (ry);

%Demodulation...........................................
 ich7 = ich6./ kmod;
 qch7 = qch6./ kmod;
 outdemodata  = qpskdemod (ich7,qch7,para,nd,m1);


function outdemodata = dwt_channel_offset_g(serdata,pa
ra,nd,m1,gilen,dctlen,sr,ebno, br,epsilon)

%serial to parallel convertion.........................
paradata = reshape(serdata,para,nd*m1);

%QPSK modulation........................................
[ich,qch] = qpskmod(paradata,para,nd,m1);

% [ich0,qch0] = compoversamp(ich01,qch01,length(ich01)
,Ipoint);
kmod = 1/sqrt(2);
ich1 = ich.*kmod;
qch1 = qch.*kmod;

%IDWT...................................................
x = ich1 + qch1.*j;
y = wavelet('D6',-1,x,'zpd');    % Invert 5 stages

% 2D wavelet transf
% R = wavelet('2D CDF 9/7',-2,Y);    % Recover X from Y
```

```
% Forward transform with 5 stages
ich2 = real (y);
qch2 = imag (y);

%Gaurd interval insertion................................
[ich3,qch3] = giins1(ich2,qch2,dctlen,gilen,nd);
dctlen2 = dctlen + gilen;

%Attenuation Calculation.................................
spow = sum(ich3.^2+qch3.^2)/nd./para;
attn = 0.5*spow*sr/br*10.^(-ebno/10);
attn = sqrt (attn);

%offset..................................................
n=para/2;
offset=exp(j*2*pi*n*epsilon/para);
i_rx_signal = ich3.*offset;
q_rx_signal = qch3.*offset;
ich3a = i_rx_signal ;
qch3a = q_rx_signal ;

%Receiever...............................................

%AWGN addition...........................................
 [ich4,qch4] = comb(ich3a,qch3a,attn);
%Guard interval removal..................................
 [ich5,qch5] = girem1 (ich4,qch4,dctlen2,gilen,nd);
%DWT.....................................................
rx = ich5 + qch5.*j;
ry =wavelet('D6',1,rx,'zpd');;  % 2D wavelet transform
%   R = wavelet('2D CDF 9/7',-2,Y); % Recover X from Y;
ich6 = real (ry);
qch6 = imag (ry);

%Demodulati..............................................
 ich7 = ich6./ kmod;
 qch7 = qch6./ kmod;
 outdemodata  = qpskdemod (ich7,qch7,para,nd,m1);


function outdemodata = fft_channel_offset_zg(serdata,p
ara,nd,m1,gilen,fftlen,sr,ebno, br,epsilon)

%serial to parallel convertion...........................
paradata = reshape(serdata,para,nd*m1);

%QPSK modulation.........................................
[ich,qch] = qpskmod(paradata,para,nd,m1);

% [ich0,qch0] = compoversamp(ich01,qch01,length(ich01)
,Ipoint);
```

```
kmod = 1/sqrt(2);
ich1 = ich.*kmod;
qch1 = qch.*kmod;

%IFFT.............................................
x = ich1 + qch1.*j;
y = ifft(x);
ich2 = real (y);
qch2 = imag (y);

%Gaurd interval insertion..........................
[ich3,qch3] = giins2(ich2,qch2,fftlen,gilen,nd);
fftlen2 = fftlen + gilen;

%Attenuation Calculation...........................

spow = sum(ich3.^2+qch3.^2)/nd./para;
attn = 0.5*spow*sr/br*10.^(-ebno/10);
attn = sqrt (attn);

%offset............................................
n=para/2;
offset=exp(j*2*pi*n*epsilon/para);
i_rx_signal = ich3.*offset;
q_rx_signal = qch3.*offset;
ich3a = i_rx_signal ;
qch3a = q_rx_signal ;

%Receiever.........................................
%AWGN addition.....................................
 [ich4,qch4] = comb(ich3a,qch3a,attn);

%Guard  interval  removal..........................
 [ich5,qch5] = girem1 (ich4,qch4,fftlen2,gilen,nd);

%FFT..............................................
rx = ich5 + qch5.*j;
ry = fft(rx);
ich6 = real (ry);
qch6 = imag (ry);

%Demodulation.....................................
 ich7 = ich6./ kmod;
 qch7 = qch6./ kmod;
 outdemodata  = qpskdemod (ich7,qch7,para,nd,m1);


function outdemodata = dct_channel_offset_zg(serdata,p
ara,nd,m1,gilen,fftlen,sr,ebno, br,epsilon)

%serial to parallel convertion.....................
paradata = reshape(serdata,para,nd*m1);
```

```
%QPSK modulation......................................
[ich,qch] = qpskmod(paradata,para,nd,m1);

% [ich0,qch0] = compoversamp(ich01,qch01,length(ich01)
,Ipoint);
kmod = 1/sqrt(2);
ich1 = ich.*kmod;
qch1 = qch.*kmod;

%IDCT.................................................
x = ich1 + qch1.*j;
y = idct(x);
ich2 = real (y);
qch2 = imag (y);

%Gaurd interval insertion.............................
[ich3,qch3] = giins2(ich2,qch2,fftlen,gilen,nd);
fftlen2 = fftlen + gilen;

%Attenuation Calculation..............................
spow = sum(ich3.^2+qch3.^2)/nd./para;
attn = 0.5*spow*sr/br*10.^(-ebno/10);
attn = sqrt (attn);

%offset...............................................
n=para/2;
offset=exp(j*2*pi*n*epsilon/para);
i_rx_signal = ich3.*offset;
q_rx_signal = qch3.*offset;
ich3a = i_rx_signal ;
qch3a = q_rx_signal ;

%Receiever............................................
%AWGN addition........................................
 [ich4,qch4] = comb(ich3a,qch3a,attn);

%Guard interval removal...............................
 [ich5,qch5] = girem1 (ich4,qch4,fftlen2,gilen,nd);

%DCT..................................................
rx = ich5 + qch5.*j;
ry = dct(rx);
ich6 = real (ry);
qch6 = imag (ry);

%Demodulation.........................................
 ich7 = ich6./ kmod;
 qch7 = qch6./ kmod;
 outdemodata  = qpskdemod (ich7,qch7,para,nd,m1);
```

```
function outdemodata = dwt_channel_offset_zg(serdata,p
ara,nd,m1,gilen,dctlen,sr,ebno, br,epsilon)

%serial to parallel convertion........................
paradata = reshape(serdata,para,nd*m1);
%QPSK modulation......................................
[ich,qch] = qpskmod(paradata,para,nd,m1);
% [ich0,qch0] = compoversamp(ich01,qch01,length(ich01)
,Ipoint);
kmod = 1/sqrt(2);
ich1 = ich.*kmod;
qch1 = qch.*kmod;
%IDWT.................................................
x = ich1 + qch1.*j;
y = wavelet('D6',-1,x,'zpd');    % Invert 5 stages

% 2D wavelet transf
% R = wavelet('2D CDF 9/7',-2,Y);    % Recover X from Y
% Forward transform with 5 stages
ich2 = real (y);
qch2 = imag (y);

%Gaurd interval insertion.............................
[ich3,qch3] = giins2(ich2,qch2,dctlen,gilen,nd);
dctlen2 = dctlen + gilen;

%Attenuation Calculation..............................

spow = sum(ich3.^2+qch3.^2)/nd./para;
attn = 0.5*spow*sr/br*10.^(-ebno/10);
attn = sqrt (attn);

%offset...............................................
n=para/2;
offset=exp(j*2*pi*n*epsilon/para);
i_rx_signal = ich3.*offset;
q_rx_signal = qch3.*offset;
ich3a = i_rx_signal ;
qch3a = q_rx_signal ;

%Receiever............................................

%AWGN addition........................................
 [ich4,qch4] = comb(ich3a,qch3a,attn);
%Guard interval removal...............................
 [ich5,qch5] = girem1 (ich4,qch4,dctlen2,gilen,nd);
%DWT..................................................
rx = ich5 + qch5.*j;
ry =wavelet('D6',1,rx,'zpd');; % 2D wavelet transform
%   R = wavelet('2D CDF 9/7',-2,Y); % Recover X from Y;
```

```
ich6 = real (ry);
qch6 = imag (ry);

%Demodulation.......................................
 ich7 = ich6./ kmod;
 qch7 = qch6./ kmod;
 outdemodata  = qpskdemod (ich7,qch7,para,nd,m1);


function outdemodata = fft_channel_estimation_no_ma
pping_g(serdata,para,nd,m1,gilen,fftlen,sr,ebno,
br,fd,flat)

paradata = reshape(serdata,para,nd*m1);
%QPSK modulation....................................
[ich,qch] = qpskmod(paradata,para,nd,m1);
% [ich0,qch0] = compoversamp(ich01,qch01,length(ich01)
,Ipoint);
kmod = 1/sqrt(2);
ich1 = ich.*kmod;
qch1 = qch.*kmod;
%channel estimation data generation..................
kndata=zeros(1,fftlen);
kndata0=2.*(rand(1,para)<0.5)-1;
kndata(1:para/2)=kndata0(1:para/2);
kndata((para/2)+1:para)=kndata0((para/2)+1:para);
ceich=kndata;
ceqch=zeros(1,para);
%data mapping.......................................
ich2=[ceich.' ich1];
qch2=[ceqch.' qch1];

%IFFT...............................................
x = ich2 + qch2.*i;
y = ifft(x);
ich3 = real (y);
qch3 = imag (y);

%Gaurd interval insertion...........................

[ich4,qch4] = giins1(ich3,qch3,fftlen,gilen,nd+1);
fftlen2 = fftlen + gilen;

%Attenuation Calculation............................

spow = sum(ich4.^2+qch4.^2)/nd./para;
attn = 0.5*spow*sr/br*10.^(-ebno/10);
attn = sqrt (attn);
%fading.............................................
%******************** Create Rayleigh fading channel
object. ********************
```

```
tstp=1/sr/(fftlen+gilen);
itau=[0,2,3,4];
dlvll=[0,10,20,25];
n0=[6,7,6,7];
th1=[0,0,0,0];
itnd1=[1000,2000,3000,4000];
now1=4;
itnd0=nd*(fftlen+gilen)*20;
[ifade,qfade,ramp,rcos,rsin]=sefade(ich4,qch4,itau,dlv
ll,th1,n0,itnd1,now1,length(ich4),tstp,fd,flat);
itnd1=itnd1+itnd0;
ich4=ifade;
qch4=qfade;

%***************** AWGN addition *******************
%Receiever........................................

%AWGN addition....................................

 [ich5,qch5] = comb(ich4,qch4,attn);
 %perfect fading compensation......................
 ifade2=1./ramp.*(rcos(1,:).*ich5+rsin(1,:).*qch5);
 qfade2=1./ramp.*(-rsin(1,:).*ich5+rcos(1,:).*qch5);
 ich5=ifade2;
 qch5=qfade2;

%Guard interval removal...........................
 [ich6,qch6] = girem1 (ich5,qch5,fftlen2,gilen,nd+1);
%FFT..............................................
rx = ich6 + qch6.*j;
ry = fft(rx);
ich7 = real (ry);
qch7 = imag (ry);
%fading compensation by channel estimation symbol.....
ce=1;
ice0=ich2(:,ce);
qce0=qch2(:,ce);
ice1=ich7(:,ce);
qce1=qch7(:,ce);
%calculate reverse rotation.......................
iv=real((1./(ice1.^2+qce1.^2)).*(ice0+i.*qce0).*(ice1-
i.*qce1));
qv=imag((1./(ice1.^2+qce1.^2)).*(ice0+i.*qce0).*(ice1-
i.*qce1));
%matrix for reverse rotation......................
ieqv1=[iv iv iv iv iv iv iv];
qeqv1=[qv qv qv qv qv qv qv];
%reverse rptation.......................
icompen=real((ich7+i.*qch7).*(ieqv1+i.*qeqv1));
qcompen=imag((ich7+i.*qch7).*(ieqv1+i.*qeqv1));
```

```
ich7=icompen;
qch7=qcompen;
%channel estimation symbol removal...................
knd=1;                  %number of known channel
                        estimation ofdm symbol
ich9=ich7(:,knd+1:nd+1);
qch9=qch7(:,knd+1:nd+1);

%Demodulation....................................
 ich10 = ich9./ kmod;
 qch10 = qch9./ kmod;
 outdemodata  = qpskdemod (ich10,qch10,para,nd,m1);


function outdemodata = dct_channel_estimation_no_ma
pping_g(serdata,para,nd,m1,gilen,fftlen,sr,ebno,
br,fd,flat)

paradata = reshape(serdata,para,nd*m1);
%QPSK modulation..................................
[ich,qch] = qpskmod(paradata,para,nd,m1);
% [ich0,qch0] = compoversamp(ich01,qch01,length(ich01)
,Ipoint);
kmod = 1/sqrt(2);
ich1 = ich.*kmod;
qch1 = qch.*kmod;
%channel estimation data generation.................
kndata=zeros(1,fftlen);
kndata0=2.*(rand(1,para)<0.5)-1;
kndata(1:para/2)=kndata0(1:para/2);
kndata((para/2)+1:para)=kndata0((para/2)+1:para);
ceich=kndata;
ceqch=zeros(1,para);
%data mapping.....................................
ich2=[ceich.' ich1];
qch2=[ceqch.' qch1];

%IDCT.............................................
x = ich2 + qch2.*i;
y = idct(x);
ich3 = real (y);
qch3 = imag (y);

%Gaurd interval insertion..........................

[ich4,qch4] = giins1(ich3,qch3,fftlen,gilen,nd+1);
fftlen2 = fftlen + gilen;

%Attenuation Calculation...........................

spow = sum(ich4.^2+qch4.^2)/nd./para;
```

```
attn = 0.5*spow*sr/br*10.^(-ebno/10);
attn = sqrt (attn);
%fading.............................................
%******************** Create Rayleigh fading channel
object. *******************
tstp=1/sr/(fftlen+gilen);
itau=[0,2,3,4];
dlvl1=[0,10,20,25];
n0=[6,7,6,7];
th1=[0,0,0,0];
itnd1=[1000,2000,3000,4000];
now1=4;
itnd0=nd*(fftlen+gilen)*20;
[ifade,qfade,ramp,rcos,rsin]=sefade(ich4,qch4,itau,dlv
l1,th1,n0,itnd1,now1,length(ich4),tstp,fd,flat);
itnd1=itnd1+itnd0;
ich4=ifade;
qch4=qfade;

%***************** AWGN addition ********************
%Receiever...........................................

%AWGN addition.......................................

 [ich5,qch5] = comb(ich4,qch4,attn);
 %perfect fading compensation.........................
 ifade2=1./ramp.*(rcos(1,:).*ich5+rsin(1,:).*qch5);
 qfade2=1./ramp.*(-rsin(1,:).*ich5+rcos(1,:).*qch5);
 ich5=ifade2;
 qch5=qfade2;

%Guard interval removal..............................
 [ich6,qch6] = girem1 (ich5,qch5,fftlen2,gilen,nd+1);
%DCT.................................................
rx = ich6 + qch6.*j;
ry = dct(rx);
ich7 = real (ry);
qch7 = imag (ry);
%fading compensation by channel estimation symbol.....
ce=1;
ice0=ich2(:,ce);
qce0=qch2(:,ce);
ice1=ich7(:,ce);
qce1=qch7(:,ce);
%calculate reverse rotation..........................
iv=real((1./(ice1.^2+qce1.^2)).*(ice0+i.*qce0).*(ice1-
i.*qce1));
qv=imag((1./(ice1.^2+qce1.^2)).*(ice0+i.*qce0).*(ice1-
i.*qce1));
%matrix for reverse rotation.........................
```

```
ieqv1=[iv iv iv iv iv iv iv];
qeqv1=[qv qv qv qv qv qv qv];
%reverse rptation......................
icompen=real((ich7+i.*qch7).*(ieqv1+i.*qeqv1));
qcompen=imag((ich7+i.*qch7).*(ieqv1+i.*qeqv1));
ich7=icompen;
qch7=qcompen;
%channel estimation symbol removal...................
knd=1;                  %number of known channel
                        estimation ofdm symbol
ich9=ich7(:,knd+1:nd+1);
qch9=qch7(:,knd+1:nd+1);

%Demodulation.....................................
 ich10 = ich9./ kmod;
 qch10 = qch9./ kmod;
 outdemodata  = qpskdemod (ich10,qch10,para,nd,m1);


function outdemodata = dwt_channel_estimation_no_ma
pping_g(serdata,para,nd,m1,gilen,fftlen,sr,ebno,
br,fd,flat)

paradata = reshape(serdata,para,nd*m1);
%QPSK modulation...................................
[ich,qch] = qpskmod(paradata,para,nd,m1);
% [ich0,qch0] = compoversamp(ich01,qch01,length(ich01)
,Ipoint);
kmod = 1/sqrt(2);
ich1 = ich.*kmod;
qch1 = qch.*kmod;
%channel estimation data generation..................
kndata=zeros(1,fftlen);
kndata0=2.*(rand(1,para)<0.5)-1;
kndata(1:para/2)=kndata0(1:para/2);
kndata((para/2)+1:para)=kndata0((para/2)+1:para);
ceich=kndata;
ceqch=zeros(1,para);
%data mapping......................................
ich2=[ceich.' ich1];
qch2=[ceqch.' qch1];
%IDWT..............................................
x = ich2 + qch2.*j;
y = wavelet('D6',-1,x,'zpd');   % Invert 5 stages
%;    % 2D wavelet transform
%   R = wavelet('2D CDF 9/7',-2,Y);   % Recover X from
Y % Forward transform with 5 stages

ich3 = real (y);
qch3 = imag (y);
```

```
%Gaurd interval insertion..............................

[ich4,qch4] = giins1(ich3,qch3,fftlen,gilen,nd+1);
fftlen2 = fftlen + gilen;

%Attenuation Calculation...............................

spow = sum(ich4.^2+qch4.^2)/nd./para;
attn = 0.5*spow*sr/br*10.^(-ebno/10);
attn = sqrt (attn);
%fading................................................
%******************** Create Rayleigh fading channel
object. ********************
tstp=1/sr/(fftlen+gilen);
itau=[0,2,3,4];
dlvl1=[0,10,20,25];
n0=[6,7,6,7];
th1=[0,0,0,0];
itnd1=[1000,2000,3000,4000];
now1=4;
itnd0=nd*(fftlen+gilen)*20;
[ifade,qfade,ramp,rcos,rsin]=sefade(ich4,qch4,itau,dlv
l1,th1,n0,itnd1,now1,length(ich4),tstp,fd,flat);
itnd1=itnd1+itnd0;
ich4=ifade;
qch4=qfade;

%***************** AWGN addition ********************
%Receiever.............................................

%AWGN addition.........................................

 [ich5,qch5] = comb(ich4,qch4,attn);
 %perfect fading compensation.........................
 ifade2=1./ramp.*(rcos(1,:).*ich5+rsin(1,:).*qch5);
 qfade2=1./ramp.*(-rsin(1,:).*ich5+rcos(1,:).*qch5);
 ich5=ifade2;
 qch5=qfade2;

%Guard interval removal................................
 [ich6,qch6] = girem1 (ich5,qch5,fftlen2,gilen,nd+1);
%DWT...................................................
rx = ich6 + qch6.*j;
ry =wavelet('D6',1,rx,'zpd');;  % 2D wavelet transform
%   R = wavelet('2D CDF 9/7',-2,Y); % Recover X from Y;
ich7 = real (ry);
qch7 = imag (ry);

%fading compensation by channel estimation symbol.....
ce=1;
ice0=ich2(:,ce);
```

```
qce0=qch2(:,ce);
ice1=ich7(:,ce);
qce1=qch7(:,ce);
%calculate reverse rotation..........................
iv=real((1./(ice1.^2+qce1.^2)).*(ice0+i.*qce0).*(ice1-
i.*qce1));
qv=imag((1./(ice1.^2+qce1.^2)).*(ice0+i.*qce0).*(ice1-
i.*qce1));
%matrix for reverse rotation.........................
ieqv1=[iv iv iv iv iv iv iv];
qeqv1=[qv qv qv qv qv qv qv];
%reverse rptation.......................
icompen=real((ich7+i.*qch7).*(ieqv1+i.*qeqv1));
qcompen=imag((ich7+i.*qch7).*(ieqv1+i.*qeqv1));
ich7=icompen;
qch7=qcompen;
%channel estimation symbol removal...................
knd=1;                %number of known channel
                      estimation ofdm symbol
ich9=ich7(:,knd+1:nd+1);
qch9=qch7(:,knd+1:nd+1);

%Demodulation........................................
 ich10 = ich9./ kmod;
 qch10 = qch9./ kmod;
 outdemodata  = qpskdemod (ich10,qch10,para,nd,m1);


function outdemodata = fft_channel_estimation_no_
mapping_zg(serdata,para,nd,m1,gilen,fftlen,sr,ebno,
br,fd,flat)

paradata = reshape(serdata,para,nd*m1);
%QPSK modulation.....................................
[ich,qch] = qpskmod(paradata,para,nd,m1);
% [ich0,qch0] = compoversamp(ich01,qch01,length(ich01)
,Ipoint);
kmod = 1/sqrt(2);
ich1 = ich.*kmod;
qch1 = qch.*kmod;
%channel estimation data generation..................
kndata=zeros(1,fftlen);
kndata0=2.*(rand(1,para)<0.5)-1;
kndata(1:para/2)=kndata0(1:para/2);
kndata((para/2)+1:para)=kndata0((para/2)+1:para);
ceich=kndata;
ceqch=zeros(1,para);
%data mapping........................................
ich2=[ceich.' ich1];
qch2=[ceqch.' qch1];
```

```
%IFFT...............................................
x = ich2 + qch2.*i;
y = ifft(x);
ich3 = real (y);
qch3 = imag (y);

%Gaurd interval insertion...........................

[ich4,qch4] = giins2(ich3,qch3,fftlen,gilen,nd+1);
fftlen2 = fftlen + gilen;

%Attenuation Calculation............................

spow = sum(ich4.^2+qch4.^2)/nd./para;
attn = 0.5*spow*sr/br*10.^(-ebno/10);
attn = sqrt (attn);
%fading.............................................
%******************** Create Rayleigh fading channel
object. *******************
tstp=1/sr/(fftlen+gilen);
itau=[0,2,3,4];
dlvl1=[0,10,20,25];
n0=[6,7,6,7];
th1=[0,0,0,0];
itnd1=[1000,2000,3000,4000];
now1=4;
itnd0=nd*(fftlen+gilen)*20;
[ifade,qfade,ramp,rcos,rsin]=sefade(ich4,qch4,itau,dlv
l1,th1,n0,itnd1,now1,length(ich4),tstp,fd,flat);
itnd1=itnd1+itnd0;
ich4=ifade;
qch4=qfade;

%**************** AWGN addition ********************
%Receiever..........................................

%AWGN addition......................................

 [ich5,qch5] = comb(ich4,qch4,attn);
 %perfect fading compensation.......................
 ifade2=1./ramp.*(rcos(1,:).*ich5+rsin(1,:).*qch5);
 qfade2=1./ramp.*(-rsin(1,:).*ich5+rcos(1,:).*qch5);
 ich5=ifade2;
 qch5=qfade2;

%Guard interval removal.............................
 [ich6,qch6] = girem1 (ich5,qch5,fftlen2,gilen,nd+1);
%FFT................................................
rx = ich6 + qch6.*j;
ry = fft(rx);
ich7 = real (ry);
qch7 = imag (ry);
```

```
%fading compensation by channel estimation symbol.....
ce=1;
ice0=ich2(:,ce);
qce0=qch2(:,ce);
ice1=ich7(:,ce);
qce1=qch7(:,ce);
%calculate reverse rotation...........................
iv=real((1./(ice1.^2+qce1.^2)).*(ice0+i.*qce0).*(ice1-
i.*qce1));
qv=imag((1./(ice1.^2+qce1.^2)).*(ice0+i.*qce0).*(ice1-
i.*qce1));
%matrix for reverse rotation..........................
ieqv1=[iv iv iv iv iv iv iv];
qeqv1=[qv qv qv qv qv qv qv];
%reverse rptation......................
icompen=real((ich7+i.*qch7).*(ieqv1+i.*qeqv1));
qcompen=imag((ich7+i.*qch7).*(ieqv1+i.*qeqv1));
ich7=icompen;
qch7=qcompen;
%channel estimation symbol removal....................
knd=1;                  %number of known channel
                        estimation ofdm symbol
ich9=ich7(:,knd+1:nd+1);
qch9=qch7(:,knd+1:nd+1);

%Demodulation.........................................
 ich10 = ich9./ kmod;
 qch10 = qch9./ kmod;
 outdemodata  = qpskdemod (ich10,qch10,para,nd,m1);


function outdemodata = dct_channel_estimation_no_
mapping_zg(serdata,para,nd,m1,gilen,fftlen,sr,ebno,
br,fd,flat)

paradata = reshape(serdata,para,nd*m1);
%QPSK modulation......................................
[ich,qch] = qpskmod(paradata,para,nd,m1);
% [ich0,qch0] = compoversamp(ich01,qch01,length(ich01)
,Ipoint);
kmod = 1/sqrt(2);
ich1 = ich.*kmod;
qch1 = qch.*kmod;
%channel estimation data generation...................
kndata=zeros(1,fftlen);
kndata0=2.*(rand(1,para)<0.5)-1;
kndata(1:para/2)=kndata0(1:para/2);
kndata((para/2)+1:para)=kndata0((para/2)+1:para);
ceich=kndata;
ceqch=zeros(1,para);
```

```
%data mapping.............................................
ich2=[ceich.' ich1];
qch2=[ceqch.' qch1];

%IDCT.....................................................
x = ich2 + qch2.*i;
y = idct(x);
ich3 = real (y);
qch3 = imag (y);

%Gaurd interval insertion.................................

[ich4,qch4] = giins2(ich3,qch3,fftlen,gilen,nd+1);
fftlen2 = fftlen + gilen;

%Attenuation Calculation..................................

spow = sum(ich4.^2+qch4.^2)/nd./para;
attn = 0.5*spow*sr/br*10.^(-ebno/10);
attn = sqrt (attn);
%fading...................................................
%******************** Create Rayleigh fading channel
object. *******************
tstp=1/sr/(fftlen+gilen);
itau=[0,2,3,4];
dlvl1=[0,10,20,25];
n0=[6,7,6,7];
th1=[0,0,0,0];
itnd1=[1000,2000,3000,4000];
now1=4;
itnd0=nd*(fftlen+gilen)*20;
[ifade,qfade,ramp,rcos,rsin]=sefade(ich4,qch4,itau,dlv
l1,th1,n0,itnd1,now1,length(ich4),tstp,fd,flat);
itnd1=itnd1+itnd0;
ich4=ifade;
qch4=qfade;

%**************** AWGN addition ********************
%Receiever................................................

%AWGN addition............................................

 [ich5,qch5] = comb(ich4,qch4,attn);
 %perfect fading compensation.............................
 ifade2=1./ramp.*(rcos(1,:).*ich5+rsin(1,:).*qch5);
 qfade2=1./ramp.*(-rsin(1,:).*ich5+rcos(1,:).*qch5);
 ich5=ifade2;
 qch5=qfade2;

%Guard interval removal...................................
 [ich6,qch6] = girem1 (ich5,qch5,fftlen2,gilen,nd+1);
```

```
%DCT.....................................
rx = ich6 + qch6.*j;
ry = dct(rx);
ich7 = real (ry);
qch7 = imag (ry);
%fading compensation by channel estimation symbol.....
ce=1;
ice0=ich2(:,ce);
qce0=qch2(:,ce);
ice1=ich7(:,ce);
qce1=qch7(:,ce);
%calculate reverse rotation...........................
iv=real((1./(ice1.^2+qce1.^2)).*(ice0+i.*qce0).*(ice1-
i.*qce1));
qv=imag((1./(ice1.^2+qce1.^2)).*(ice0+i.*qce0).*(ice1-
i.*qce1));
%matrix for reverse rotation..........................
ieqv1=[iv iv iv iv iv iv iv];
qeqv1=[qv qv qv qv qv qv qv];
%reverse rptation......................
icompen=real((ich7+i.*qch7).*(ieqv1+i.*qeqv1));
qcompen=imag((ich7+i.*qch7).*(ieqv1+i.*qeqv1));
ich7=icompen;
qch7=qcompen;
%channel estimation symbol removal...................
knd=1;                 %number of known channel
                       estimation ofdm symbol
ich9=ich7(:,knd+1:nd+1);
qch9=qch7(:,knd+1:nd+1);

%Demodulation.................................
 ich10 = ich9./ kmod;
 qch10 = qch9./ kmod;
 outdemodata  = qpskdemod (ich10,qch10,para,nd,m1);


function outdemodata = dwt_channel_estimation_no_
mapping_zg(serdata,para,nd,m1,gilen,fftlen,sr,ebno,
br,fd,flat)

paradata = reshape(serdata,para,nd*m1);
%QPSK modulation.................................
[ich,qch] = qpskmod(paradata,para,nd,m1);
% [ich0,qch0] = compoversamp(ich01,qch01,length(ich01)
,Ipoint);
kmod = 1/sqrt(2);
ich1 = ich.*kmod;
qch1 = qch.*kmod;
%channel estimation data generation..................
kndata=zeros(1,fftlen);
kndata0=2.*(rand(1,para)<0.5)-1;
```

```
kndata(1:para/2)=kndata0(1:para/2);
kndata((para/2)+1:para)=kndata0((para/2)+1:para);
ceich=kndata;
ceqch=zeros(1,para);
%data mapping....................................
ich2=[ceich.' ich1];
qch2=[ceqch.' qch1];

%IDWT............................................
x = ich2 + qch2.*j;
y = wavelet('D6',-1,x,'zpd');    % Invert 5 stages
%;    % 2D wavelet transform
%   R = wavelet('2D CDF 9/7',-2,Y);   % Recover X from
Y % Forward transform with 5 stages

ich3 = real (y);
qch3 = imag (y);

%Gaurd interval insertion.........................

[ich4,qch4] = giins2(ich3,qch3,fftlen,gilen,nd+1);
fftlen2 = fftlen + gilen;

%Attenuation Calculation..........................

spow = sum(ich4.^2+qch4.^2)/nd./para;
attn = 0.5*spow*sr/br*10.^(-ebno/10);
attn = sqrt (attn);
%fading...........................................
%******************** Create Rayleigh fading channel
object. *******************
tstp=1/sr/(fftlen+gilen);
itau=[0,2,3,4];
dlvl1=[0,10,20,25];
n0=[6,7,6,7];
th1=[0,0,0,0];
itnd1=[1000,2000,3000,4000];
now1=4;
itnd0=nd*(fftlen+gilen)*20;
[ifade,qfade,ramp,rcos,rsin]=sefade(ich4,qch4,itau,dlv
l1,th1,n0,itnd1,now1,length(ich4),tstp,fd,flat);
itnd1=itnd1+itnd0;
ich4=ifade;
qch4=qfade;


%***************** AWGN addition ********************
%Receiever........................................

%AWGN addition....................................

 [ich5,qch5] = comb(ich4,qch4,attn);
```

```
 %perfect fading compensation..........................
 ifade2=1./ramp.*(rcos(1,:).*ich5+rsin(1,:).*qch5);
 qfade2=1./ramp.*(-rsin(1,:).*ich5+rcos(1,:).*qch5);
 ich5=ifade2;
 qch5=qfade2;

%Guard interval removal..............................
 [ich6,qch6] = girem1 (ich5,qch5,fftlen2,gilen,nd+1);
%DWT..................................................
rx = ich6 + qch6.*j;
ry =wavelet('D6',1,rx,'zpd');;  % 2D wavelet transform
%   R = wavelet('2D CDF 9/7',-2,Y); % Recover X from Y;
ich7 = real (ry);
qch7 = imag (ry);

%fading compensation by channel estimation symbol.....
ce=1;
ice0=ich2(:,ce);
qce0=qch2(:,ce);
ice1=ich7(:,ce);
qce1=qch7(:,ce);
%calculate reverse rotation...........................
iv=real((1./(ice1.^2+qce1.^2)).*(ice0+i.*qce0).*(ice1-
i.*qce1));
qv=imag((1./(ice1.^2+qce1.^2)).*(ice0+i.*qce0).*(ice1-
i.*qce1));
%matrix for reverse rotation..........................
ieqv1=[iv iv iv iv iv iv iv];
qeqv1=[qv qv qv qv qv qv qv];
%reverse rptation......................
icompen=real((ich7+i.*qch7).*(ieqv1+i.*qeqv1));
qcompen=imag((ich7+i.*qch7).*(ieqv1+i.*qeqv1));
ich7=icompen;
qch7=qcompen;
%channel estimation symbol removal...................
knd=1;              %number of known channel
                    estimation ofdm symbol
ich9=ich7(:,knd+1:nd+1);
qch9=qch7(:,knd+1:nd+1);

%Demodulation.........................................
 ich10 = ich9./ kmod;
 qch10 = qch9./ kmod;
 outdemodata  = qpskdemod (ich10,qch10,para,nd,m1);


function [iout,qout] = qpskmod(paradata,para,nd,m1)

%****************  variables *********************

% paradata : input data (para-by-nd matrix)
% iout     : output Ich data
```

```
% qout       : output Qch data
% para       : Number of parallel channels
% nd         : Number of data
% m1         : Number of modulation levels
%              (QPSK = 2  16QAM = 4)

%*******************************************************

m2 = m1./2;

paradata2 = paradata.*2-1;
count2    = 0;

for  jjj = 1: nd
        isi = zeros (para,1);
        isq = zeros (para,1);

    for ii = 1 : m2
        isi =  isi + 2.^(m2 - ii).*paradata2
        ((1:para),ii+count2) ;
        isq =  isq + 2.^(m2 - ii).*paradata2
        ((1:para),m2+ii+count2) ;
    end

        iout((1:para),jjj) = isi;
        qout((1:para),jjj) = isq;

        count2 = count2 + m1;

end
%****************** End of file *********************


function [demodata] = qpskdemod( idata, qdata
,para,nd,m1)

%****************  variables *********************

% idata     : input Ich data
% qdata     : input Qch data
% demodata : demodulated data
% para      : Number of parallel channels
% nd        : Number of data
% m1        : Number of modulation levels
%             (QPSK = 2  16QAM = 4)

%*******************************************************

demodata = zeros (para,m1*nd);
demodata ((1:para) , ( 1: m1 :m1*nd-1)) =
idata((1:para), (1:nd)) >= 0;
demodata ((1:para) , ( 2: m1 :m1*nd)) =
qdata((1:para), (1:nd)) >= 0;
```

```
%***************** End of file ********************

function varargout = wavelet(WaveletName,Level,
X,Ext,Dim)

%WAVELET  Discrete wavelet transform.
%  Y = WAVELET(W,L,X) computes the L-stage discrete
   wavelet transform
%  (DWT) of signal X using wavelet W.  The length of X
   must be
%  divisible by 2^L.  For the inverse transform,
   WAVELET(W,-L,X)
%   inverts L stages.  Choices for W are
%   'Haar'                                  Haar
%   'D1','D2','D3','D4','D5','D6'           Daubechies'
%   'Sym1','Sym2','Sym3','Sym4','Sym5',     Symlets
    'Sym6'
%   'Coif1','Coif2'                         Coiflets
%   'BCoif1'                                Coiflet-like[2]
%   'Spline Nr.Nd'(or 'bior Nr.Nd')for      Splines
%     Nr = 0, Nd = 0,1,2,3,4,5,6,7 or 8
%     Nr = 1, Nd = 0,1,3,5, or 7
%     Nr = 2, Nd = 0,1,2,4,6, or 8
%     Nr = 3, Nd = 0,1,3,5, or 7
%     Nr = 4, Nd = 0,1,2,4,6, or 8
%     Nr = 5, Nd = 0,1,3, or 5
%   'RSpline Nr.Nd' for the same Nr.Nd      Reverse splines
    pairs
%   'S+P (2,2)','S+P (4,2)','S+P (6,2)',    S+P wavelets[3]
%   'S+P (4,4)','S+P (2+2,2)'
%   'TT'                                    "Two-Ten" [5]
%   'LC 5/3','LC 2/6','LC 9/7-M',           Low Complexity
    'LC 2/10',                              [1]
%     'LC 5/11-C','LC 5/11-A','LC 6/14',
%     'LC 13/7-T','LC 13/7-C'
%     'Le Gall 5/3','CDF 9/7'                 JPEG2000 [7]
%     'V9/3'                                  Visual [8]
%     'Lazy'                                  Lazy wavelet
%  Case and spaces are ignored in wavelet names, for
   example, 'Sym4'
%  may also be written as 'sym 4'.  Some wavelets have
   multiple names,
%  'D1', 'Sym1', and 'Spline 1.1' are aliases of the
   Haar wavelet.
%
%  WAVELET(W) displays information about wavelet W and
   plots the
%  primal and dual scaling and wavelet functions.
%
```

```
%  For 2D transforms, prefix W with '2D'.  For example,
   '2D S+P (2,2)'
%  specifies a 2D (tensor) transform with the S+P (2,2)
   wavelet.
%  2D transforms require that X is either MxN or MxNxP
   where M and N
%  are divisible by 2^L.
%
%  WAVELET(W,L,X,EXT) specifies boundary handling EXT.
   Choices are
%   'sym'    Symmetric extension (same as 'wsws')
%   'asym'   Antisymmetric extension, whole-point
             antisymmetry
%   'zpd'    Zero-padding
%   'per'    Periodic extension
%   'sp0'    Constant extrapolation
%
%  Various symmetric extensions are supported:
%   'wsws'   Whole-point symmetry (WS) on both boundaries
%   'hshs'   Half-point symmetry (HS) on both boundaries
%   'wshs'   WS left boundary, HS right boundary
%   'hsws'   HS left boundary, WS right boundary
%
%  Antisymmetric boundary handling is used by default,
   EXT = 'asym'.
%
%  WAVELET(...,DIM) operates along dimension DIM.
%
%  [H1,G1,H2,G2] = WAVELET(W,'filters') returns the filters
%  associated with wavelet transform W.  Each filter is
   represented
%  by a cell array where the first cell contains an
   array of
%  coefficients and the second cell contains a scalar
   of the leading
%  Z-power.
%
%  [X,PHI1] = WAVELET(W,'phi1') returns an
   approximation of the
%  scaling function associated with wavelet transform W.
%  [X,PHI1] = WAVELET(W,'phi1',N) approximates the
   scaling function
%  with resolution 2^-N.  Similarly,
%  [X,PSI1] = WAVELET(W,'psi1',...),
%  [X,PHI2] = WAVELET(W,'phi2',...),
%  and [X,PSI2] = WAVELET(W,'psi2',...) return
   approximations of the
%  wavelet function, dual scaling function, and dual
   wavelet function.
%
```

```
% Wavelet transforms are implemented using the
  lifting scheme [4].
% For general background on wavelets, see for example
  [6].
%
%
% Examples:
% % Display information about the S+P (4,4) wavelet
% wavelet('S+P (4,4)');
%
% % Plot a wavelet decomposition
% t = linspace(0,1,256);
% X = exp(-t) + sqrt(t - 0.3).*(t > 0.3) - 0.2*(t > 0.6);
% wavelet('RSpline 3.1',3,X);        % Plot the
  decomposition of X
%
% % Sym4 with periodic boundaries
% Y = wavelet('Sym4',5,X,'per');    % Forward
  transform with 5 stages
% R = wavelet('Sym4',-5,Y,'per');   % Invert 5 stages
%
% % 2D transform on an image
% t = linspace(-1,1,128); [x,y] = meshgrid(t,t);
% X = ((x+1).*(x-1) - (y+1).*(y-1)) + real(sqrt(0.4 -
  x.^2 - y.^2));
% Y = wavelet('2D CDF 9/7',2,X);   % 2D wavelet
  transform
% R = wavelet('2D CDF 9/7',-2,Y);   % Recover X from Y
% imagesc(abs(Y).^0.2); colormap(gray); axis image;
%
% % Plot the Daubechies 2 scaling function
% [x,phi] = wavelet('D2','phi');
% plot(x,phi);
%
% References:
% [1] M. Adams and F. Kossentini. June,2000. Reversible
      integer-to-integer wavelet transforms for image
      compression. IEEE Trans. on Image Proc. 9(6).
% [2] M. Antonini, M. Barlaud, P. Mathieu and I.
      Daubechies. 1992. Image coding using wavelet
      transforms. IEEE Trans. Image Processing 1: 205-
      220.
% [3] R. Calderbank, I. Daubechies, W. Sweldens and
      Boon-Lock Yeo. 1997. Lossless image compression
      using integer to integer wavelet transforms. ICIP
      IEEE Press 1: 596-599.
% [4] I. Daubechies and W. Sweldens. 1996. Factoring
      Wavelet Transforms into Lifting Steps.
% [5] D. Le Gall and A. Tabatabai. 1988. Subband coding
      of digital images using symmetric short kernel
```

```
        filters and arithmetic coding techniques. ICASSP'
        88: 761-765.
%   [6] S. Mallat. 1999. A Wavelet Tour of Signal Processing.
        Academic Press.
%   [7] M. Unser and T. Blu. September, 2003. Mathematical
        properties of the JPEG2000 wavelet filters. IEEE
        Trans. on Image Proc. 12(9).
%   [8] Qinghai Wang and Yulong Mo. December, 2004. Choice
        of wavelet base in JPEG2000. Computer Engineering
        30(23).
% Pascal Getreuer 2005-2006

if nargin < 1, error('Not enough input arguments.'); end
if ~ischar(WaveletName), error('Invalid wavelet name.');
end

% Get a lifting scheme sequence for the specified wavelet
Flag1D = isempty(findstr(lower(WaveletName),'2d'));
[Seq,ScaleS,ScaleD,Family] = getwavelet(WaveletName);

if isempty(Seq)
   error(['Unknown wavelet, ''',WaveletName,'''.']);
end

if nargin < 2, Level = ''; end
if ischar(Level)
   [h1,g1] = seq2hg(Seq,ScaleS,ScaleD,0);
   [h2,g2] = seq2hg(Seq,ScaleS,ScaleD,1);

   if strcmpi(Level,'filters')
      varargout = {h1,g1,h2,g2};
   else
      if nargin < 3, X = 6; end

      switch lower(Level)
      case {'phi1','phi'}
         [x1,phi] = cascade(h1,g1,pow2(-X));
         varargout = {x1,phi};
      case {'psi1','psi'}
         [x1,phi,x2,psi] = cascade(h1,g1,pow2(-X));
         varargout = {x2,psi};
      case 'phi2'
         [x1,phi] = cascade(h2,g2,pow2(-X));
         varargout = {x1,phi};
      case 'psi2'
         [x1,phi,x2,psi] = cascade(h2,g2,pow2(-X));
         varargout = {x2,psi};
      case ''
         fprintf('\n%s wavelet ''%s''
         ',Family,WaveletName);
```

```
        if all(abs([norm(h1{1}),norm(h2{1})] - 1) <
        1e-11)
           fprintf('(orthogonal)\n');
        else
           fprintf('(biorthogonal)\n');
        end

        fprintf('Vanishing moments: %d analysis, %d
        reconstruction\n',...
           numvanish(g1{1}),numvanish(g2{1}));
        fprintf('Filter lengths: %d/%d-tap\n',...
           length(h1{1}),length(g1{1}));
        fprintf('Implementation lifting steps:
        %d\n\n',...
           size(Seq,1)-all([Seq{1,:}] == 0));

        fprintf('h1(z) = %s\n',filterstr(h1,ScaleS));
        fprintf('g1(z) = %s\n',filterstr(g1,ScaleD));
        fprintf('h2(z) = %s\n',filterstr(h2,1/ScaleS));
        fprintf('g2(z) = %s\n\n',filterstr(g2,1/ScaleD));

        [x1,phi,x2,psi] = cascade(h1,g1,pow2(-X));
        subplot(2,2,1);
        plot(x1,phi,'b-');
        if diff(x1([1,end]))>0,xlim(x1([1,end])); end
        title('\phi_1');
        subplot(2,2,3);
        plot(x2,psi,'b-');
        if diff(x2([1,end]))>0,xlim(x2([1,end])); end
        title('\psi_1');
        [x1,phi,x2,psi] = cascade(h2,g2,pow2(-X));
        subplot(2,2,2);
        plot(x1,phi,'b-');
        if diff(x1([1,end]))>0,xlim(x1([1,end])); end
        title('\phi_2');
        subplot(2,2,4);
        plot(x2,psi,'b-');
        if diff(x2([1,end]))>0,xlim(x2([1,end])); end
        title('\psi_2');
        set(gcf,'NextPlot','replacechildren');
     otherwise
        error(['Invalid parameter, ''',Level,'''.']);
     end
  end

  return;
elseif nargin < 5
  % Use antisymmetric extension by default
  if nargin < 4
     if nargin < 3, error('Not enough input
     arguments.'); end
```

```
      Ext = 'asym';
   end

   Dim = min(find(size(X) ~= 1));
   if isempty(Dim), Dim = 1; end
end

if any(size(Level) ~= 1), error('Invalid decomposition
level.'); end

NumStages = size(Seq,1);
EvenStages = ~rem(NumStages,2);

if Flag1D   % 1D Transfrom
   %%% Convert N-D array to a 2-D array with dimension
   Dim along the columns %%%
   XSize = size(X);    % Save original dimensions
   N = XSize(Dim);
   M = prod(XSize)/N;
   Perm = [Dim:max(length(XSize),Dim),1:Dim-1];
   X = double(reshape(permute(X,Perm),N,M));

   if M == 1 & nargout == 0 & Level > 0
      % Create a figure of the wavelet decomposition
      set(gcf,'NextPlot','replace');
      subplot(Level+2,1,1);
      plot(X);
      title('Wavelet Decomposition');
      axis tight; axis off;

      X = feval(mfilename,WaveletName,Level,X,Ext,1);

      for i = 1:Level
         N2 = N;
         N = 0.5*N;
         subplot(Level+2,1,i+1);
         a = max(abs(X(N+1:N2)))*1.1;
         plot(N+1:N2,X(N+1:N2),'b-');
         ylabel(['d',sprintf('_%c',num2str(i))]);
         axis([N+1,N2,-a,a]);
      end

      subplot(Level+2,1,Level+2);
      plot(X(1:N),'-');
      xlabel('Coefficient Index');
      ylabel('s_1');
      axis tight;
      set(gcf,'NextPlot','replacechildren');
      varargout = {X};
      return;
   end
```

```
if rem(N,pow2(abs(Level))), error('Signal length
must be divisible by 2^L.'); end
if N < pow2(abs(Level)), error('Signal length too
small for transform level.'); end

if Level >= 0             % Forward transform
   for i = 1:Level
      Xo = X(2:2:N,:);
      Xe = X(1:2:N,:) + xfir(Seq{1,1},Seq{1,2},Xo,Ext);

      for k = 3:2:NumStages
         Xo = Xo + xfir(Seq{k-1,1},Seq{k-
         1,2},Xe,Ext);
         Xe = Xe + xfir(Seq{k,1},Seq{k,2},Xo,Ext);
         end

      if EvenStages
         Xo = Xo + xfir(Seq{NumStages,1},Seq{NumStag
         es,2},Xe,Ext);
      end

      X(1:N,:) = [Xe*ScaleS; Xo*ScaleD];
      N = 0.5*N;
   end
else                        % Inverse transform
   N = N * pow2(Level);

   for i = 1:-Level
      N2 = 2*N;
      Xe = X(1:N,:)/ScaleS;
      Xo = X(N+1:N2,:)/ScaleD;

      if EvenStages
         Xo = Xo - xfir(Seq{NumStages,1},Seq{NumStag
         es,2},Xe,Ext);
      end

      for k = NumStages - EvenStages:-2:3
         Xe = Xe - xfir(Seq{k,1},Seq{k,2},Xo,Ext);
         Xo = Xo - xfir(Seq{k-1,1},Seq{k-
         1,2},Xe,Ext);
      end

      X([1:2:N2,2:2:N2],:) = [Xe -
      xfir(Seq{1,1},Seq{1,2},Xo,Ext); Xo];
      N = N2;
   end
end

X = ipermute(reshape(X,XSize(Perm)),Perm); % Restore
original array dimensions
```

```
else         % 2D Transfrom
   N = size(X);

   if length(N) > 3 | any(rem(N([1,2]),pow2
abs(Level))))
      error('Input size must be either MxN or MxNxP
where M and N are divisible by 2^L.');
   end

   if Level >= 0   % 2D Forward transform
      for i = 1:Level
         Xo = X(2:2:N(1),1:N(2),:);
         Xe = X(1:2:N(1),1:N(2),:) +
         xfir(Seq{1,1},Seq{1,2},Xo,Ext);

         for k = 3:2:NumStages
            Xo = Xo + xfir(Seq{k-1,1},Seq{k-
            1,2},Xe,Ext);
            Xe = Xe + xfir(Seq{k,1},Seq{k,2},Xo,Ext);
         end

         if EvenStages
            Xo = Xo + xfir(Seq{NumStages,1},Seq{NumStag
            es,2},Xe,Ext);
         end

         X(1:N(1),1:N(2),:) = [Xe*ScaleS; Xo*ScaleD];

         Xo = permute(X(1:N(1),2:2:N(2),:),[2,1,3]);
         Xe = permute(X(1:N(1),1:2:N(2),:),[2,1,3]) ...
            + xfir(Seq{1,1},Seq{1,2},Xo,Ext);

         for k = 3:2:NumStages
            Xo = Xo + xfir(Seq{k-1,1},Seq{k-
            1,2},Xe,Ext);
            Xe = Xe + xfir(Seq{k,1},Seq{k,2},Xo,Ext);
         end

         if EvenStages
            Xo = Xo + xfir(Seq{NumStages,1},Seq{NumStag
            es,2},Xe,Ext);
         end

         X(1:N(1),1:N(2),:) =
         [permute(Xe,[2,1,3])*ScaleS,...
               permute(Xo,[2,1,3])*ScaleD];
         N = 0.5*N;
      end
   else              % 2D Inverse transform
      N = N*pow2(Level);
```

```
        for i = 1:-Level
           N2 = 2*N;
           Xe = permute(X(1:N2(1),1:N(2),:),[2,1,3])/
           ScaleS;
           Xo = permute(X(1:N2(1),N(2)+1
           :N2(2),:),[2,1,3])/ScaleD;

           if EvenStages
              Xo = Xo - xfir(Seq{NumStages,1},Seq{NumSta
              ges,2},Xe,Ext);
           end

           for k = NumStages - EvenStages:-2:3
              Xe = Xe - xfir(Seq{k,1},Seq{k,2},Xo,Ext);
              Xo = Xo - xfir(Seq{k-1,1},Seq{k-
              1,2},Xe,Ext);
           end

           X(1:N2(1),[1:2:N2(2),2:2:N2(2)],:) = ...
              [permute(Xe - xfir(Seq{1,1},Seq{1,2},Xo,E
              xt),[2,1,3]), ...
                 permute(Xo,[2,1,3])];

           Xe = X(1:N(1),1:N2(2),:)/ScaleS;
           Xo = X(N(1)+1:N2(1),1:N2(2),:)/ScaleD;

           if EvenStages
              Xo = Xo - xfir(Seq{NumStages,1},Seq{NumStag
              es,2},Xe,Ext);
           end

           for k = NumStages - EvenStages:-2:3
              Xe = Xe - xfir(Seq{k,1},Seq{k,2},Xo,Ext);
              Xo = Xo - xfir(Seq{k-1,1},Seq{k-
              1,2},Xe,Ext);
           end

           X([1:2:N2(1),2:2:N2(1)],1:N2(2),:) = ...
              [Xe - xfir(Seq{1,1},Seq{1,2},Xo,Ext); Xo];
           N = N2;
        end
    end
end

varargout{1} = X;
return;


function [Seq,ScaleS,ScaleD,Family] =
getwavelet(WaveletName)
%GETWAVELET   Get wavelet lifting scheme sequence.
% Pascal Getreuer 2005-2006
```

```
WaveletName = strrep(WaveletName,'bior','spline');
ScaleS = 1/sqrt(2);
ScaleD = 1/sqrt(2);
Family = 'Spline';

switch strrep(strrep(lower(WaveletName),'2d',''),'
',''')
case {'haar','d1','db1','sym1','spline1.1','rspli
ne1.1'}
   Seq = {1,0;-0.5,0};
   ScaleD = -sqrt(2);
   Family = 'Haar';
case {'d2','db2','sym2'}
   Seq = {sqrt(3),0;[-sqrt(3),2-sqrt(3)]/4,0;-1,1};
   ScaleS = (sqrt(3)-1)/sqrt(2);
   ScaleD = (sqrt(3)+1)/sqrt(2);
   Family = 'Daubechies';
case {'d3','db3','sym3'}
   Seq = {2.4254972439123361,0;[-
   0.3523876576801823,0.0793394561587384],0;
   [0.5614149091879961,-2.8953474543648969],2;-
   0.0197505292372931,-2};
   ScaleS = 0.4318799914853075;
   ScaleD = 2.3154580432421348;
   Family = 'Daubechies';
case {'d4','db4'}
   Seq = {0.3222758879971411,-
   1;[0.3001422587485443,1.1171236051605939],1;
   [-0.1176480867984784,0.0188083527262439],-1;
   [-0.6364282711906594,-2.1318167127552199],1;
   [0.0247912381571950,-0.1400392377326117,
   0.4690834789110281],2};
   ScaleS = 1.3621667200737697;
   ScaleD = 0.7341245276832514;
   Family = 'Daubechies';
case {'d5','db5'}
   Seq = {0.2651451428113514,-1;[-0.2477292913288009,-
   0.9940591341382633],1;[-0.2132742982207803,
   0.5341246460905558],1;[0.7168557197126235,-
   0.2247352231444452],-1;[-0.0121321866213973,
   0.0775533344610336],3;0.035764924629411,-3};
   ScaleS = 1.3101844387211246;
   ScaleD = 0.7632513182465389;
   Family = 'Daubechies';
case {'d6','db6'}
   Seq = {4.4344683000391223,0;
   [-0.214593449940913,0.0633131925095066],0;
   [4.4931131753641633,-9.970015617571832],2;
   [-0.0574139367993266,0.0236634936395882],-2;
   [0.6787843541162683,-2.3564970162896977],4;
```

```
        [-0.0071835631074942,0.0009911655293238],
        -4;-0.0941066741175849,5};
        ScaleS = 0.3203624223883869;
        ScaleD = 3.1214647228121661;
        Family = 'Daubechies';
    case 'sym4'
        Seq = {-0.3911469419700402,0;[0.3392439918649451,
        0.1243902829333865],0;[-0.1620314520393038,
        1.4195148522334731],1;-[0.1459830772565225,
        0.4312834159749964],1;1.049255198049293,-1};
        ScaleS = 0.6366587855802818;
        ScaleD = 1.5707000714496564;
        Family = 'Symlet';
    case 'sym5'
        Seq = {0.9259329171294208,0;-
        [0.1319230270282341,0.4985231842281166],1;
        [1.452118924420613,0.4293261204657586],0;
        [-0.2804023843755281,0.0948300395515551],0;
        -[0.7680659387165244,1.9589167118877153],
        1;0.1726400850543451,0};
        ScaleS = 0.4914339446751972;
        ScaleD = 2.0348614718930915;
        Family = 'Symlet';
    case 'sym6'
        Seq = {-0.2266091476053614,0;[0.2155407618197651,-
        1.2670686037583443],0;[-4.2551584226048398,
        0.5047757263881194],2;[0.2331599353469357,
        0.0447459687134724],-2;[6.6244572505007815,
        -18.389000853969371],4;[-0.0567684937266291,
        0.1443950619899142],-4;-5.5119344180654508,5};
        ScaleS = -0.5985483742581210;
        ScaleD = -1.6707087396895259;
        Family = 'Symlet';
    case 'coif1'
        Seq = {-4.6457513110481772,0;[0.205718913884,0.1171
        567416519999],0;[0.6076252184992341,
        -7.468626966435207],2;-0.0728756555332089,-2};
        ScaleS = -0.5818609561112537;
        ScaleD = -1.7186236496830642;
        Family = 'Coiflet';
    case 'coif2'
        Seq = {-2.5303036209828274,0;[0.3418203790296641,-
        0.2401406244344829],0;[15.268378737252995,
        3.1631993897610227],2;[-0.0646171619180252,
        0.005717132970962],-2;[13.59117256930759,
        -63.95104824798802],4;[-0.0018667030862775,
        0.0005087264425263],-4;-3.7930423341992774,5};
        ScaleS = 0.1076673102965570;
        ScaleD = 9.2878701738310099;
        Family = 'Coiflet';
```

```
case 'bcoif1'
   Seq = {0,0;-[1,1]/5,1;[5,5]/14,0;-[21,21]/100,1};
   ScaleS = sqrt(2)*7/10;
   ScaleD = sqrt(2)*5/7;
   Family = 'Nearly orthonormal Coiflet-like';
case {'lazy','spline0.0','rspline0.0','d0'}
   Seq = {0,0};
   ScaleS = 1;
   ScaleD = 1;
   Family = 'Lazy';
case {'spline0.1','rspline0.1'}
   Seq = {1,-1};
   ScaleD = 1;
case {'spline0.2','rspline0.2'}
   Seq = {[1,1]/2,0};
   ScaleD = 1;
case {'spline0.3','rspline0.3'}
   Seq = {[-1,6,3]/8,1};
   ScaleD = 1;
case {'spline0.4','rspline0.4'}
   Seq = {[-1,9,9,-1]/16,1};
   ScaleD = 1;
case {'spline0.5','rspline0.5'}
   Seq = {[3,-20,90,60,-5]/128,2};
   ScaleD = 1;
case {'spline0.6','rspline0.6'}
   Seq = {[3,-25,150,150,-25,3]/256,2};
   ScaleD = 1;
case {'spline0.7','rspline0.7'}
   Seq = {[-5,42,-175,700,525,-70,7]/1024,3};
   ScaleD = 1;
case {'spline0.8','rspline0.8'}
   Seq = {[-5,49,-245,1225,1225,-245,49,-5]/2048,3};
   ScaleD = 1;
case {'spline1.0','rspline1.0'}
   Seq = {0,0;-1,0};
   ScaleS = sqrt(2);
   ScaleD = -1/sqrt(2);
case {'spline1.3','rspline1.3'}
   Seq = {0,0;-1,0;[-1,8,1]/16,1};
   ScaleS = sqrt(2);
   ScaleD = -1/sqrt(2);
case {'spline1.5','rspline1.5'}
   Seq = {0,0;-1,0;[3,-22,128,22,-3]/256,2};
   ScaleS = sqrt(2);
   ScaleD = -1/sqrt(2);
case {'spline1.7','rspline1.7'}
   Seq = {0,0;-1,0;[-5,44,-201,1024,201,-
   44,5]/2048,3};
   ScaleS = sqrt(2);
```

```
   ScaleD = -1/sqrt(2);
case {'spline2.0','rspline2.0'}
   Seq = {0,0;-[1,1]/2,1};
   ScaleS = sqrt(2);
   ScaleD = 1;
case {'spline2.1','rspline2.1'}
   Seq = {0,0;-[1,1]/2,1;0.5,0};
   ScaleS = sqrt(2);
case {'spline2.2','rspline2.2','cdf5/3','legall5/3','
s+p(2,2)','lc5/3'}
   Seq = {0,0;-[1,1]/2,1;[1,1]/4,0};
   ScaleS = sqrt(2);
case {'spline2.4','rspline2.4'}
   Seq = {0,0;-[1,1]/2,1;[-3,19,19,-3]/64,1};
   ScaleS = sqrt(2);
case {'spline2.6','rspline2.6'}
   Seq = {0,0;-[1,1]/2,1;[5,-39,162,162,-39,5]/512,2};
   ScaleS = sqrt(2);
case {'spline2.8','rspline2.8'}
   Seq = {0,0;-[1,1]/2,1;[-35,335,-1563,5359,5359,-
   1563,335,-35]/16384,3};
   ScaleS = sqrt(2);
case {'spline3.0','rspline3.0'}
   Seq = {-1/3,-1;-[3,9]/8,1};
   ScaleS = 3/sqrt(2);
   ScaleD = 2/3;
case {'spline3.1','rspline3.1'}
   Seq = {-1/3,-1;-[3,9]/8,1;4/9,0};
   ScaleS = 3/sqrt(2);
   ScaleD = -2/3;
case {'spline3.3','rspline3.3'}
   Seq = {-1/3,-1;-[3,9]/8,1;[-3,16,3]/36,1};
   ScaleS = 3/sqrt(2);
   ScaleD = -2/3;
case {'spline3.5','rspline3.5'}
   Seq = {-1/3,-1;-[3,9]/8,1;[5,-34,128,34,-5]/288,2};
   ScaleS = 3/sqrt(2);
   ScaleD = -2/3;
case {'spline3.7','rspline3.7'}
   Seq = {-1/3,-1;-[3,9]/8,1;[-35,300,-
   1263,4096,1263,-300,35]/9216,3};
   ScaleS = 3/sqrt(2);
   ScaleD = -2/3;
case {'spline4.0','rspline4.0'}
   Seq = {-[1,1]/4,0;-[1,1],1};
   ScaleS = 4/sqrt(2);
   ScaleD = 1/sqrt(2);
   ScaleS = 1; ScaleD = 1;
case {'spline4.1','rspline4.1'}
   Seq = {-[1,1]/4,0;-[1,1],1;6/16,0};
```

```
   ScaleS = 4/sqrt(2);
   ScaleD = 1/2;
case {'spline4.2','rspline4.2'}
   Seq = {-[1,1]/4,0;-[1,1],1;[3,3]/16,0};
   ScaleS = 4/sqrt(2);
   ScaleD = 1/2;
case {'spline4.4','rspline4.4'}
   Seq = {-[1,1]/4,0;-[1,1],1;[-5,29,29,-5]/128,1};
   ScaleS = 4/sqrt(2);
   ScaleD = 1/2;
case {'spline4.6','rspline4.6'}
   Seq = {-[1,1]/4,0;-[1,1],1;[35,-265,998,998,-
   265,35]/4096,2};
   ScaleS = 4/sqrt(2);
   ScaleD = 1/2;
case {'spline4.8','rspline4.8'}
   Seq = {-[1,1]/4,0;-[1,1],1;[-63,595,-
   2687,8299,8299,-2687,595,-63]/32768,3};
   ScaleS = 4/sqrt(2);
   ScaleD = 1/2;
case {'spline5.0','rspline5.0'}
   Seq = {0,0;-1/5,0;-[5,15]/24,0;-[9,15]/10,1};
   ScaleS = 3*sqrt(2);
   ScaleD = sqrt(2)/6;
case {'spline5.1','rspline5.1'}
   Seq = {0,0;-1/5,0;-[5,15]/24,0;-[9,15]/10,1;1/3,0};
   ScaleS = 3*sqrt(2);
   ScaleD = sqrt(2)/6;
case {'spline5.3','rspline5.3'}
   Seq = {0,0;-1/5,0;-[5,15]/24,0;-[9,15]/10,1;[-
   5,24,5]/72,1};
   ScaleS = 3*sqrt(2);
   ScaleD = sqrt(2)/6;
case {'spline5.5','rspline5.5'}
   Seq = {0,0;-1/5,0;-[5,15]/24,0;-[9,15]/10,1;[35,-
   230,768,230,-35]/2304,2};
   ScaleS = 3*sqrt(2);
   ScaleD = sqrt(2)/6;
case {'cdf9/7'}
   Seq = {0,0;[1,1]*-1.5861343420693648,1;[1,1]*-
   0.0529801185718856,0;[1,1]*0.8829110755411875,
   1;[1,1]*0.4435068520511142,0};
   ScaleS = 1.1496043988602418;
   ScaleD = 1/ScaleS;
   Family = 'Cohen-Daubechies-Feauveau';
case 'v9/3'
   Seq = {0,0;[-1,-1]/2,1;[1,19,19,1]/80,1};
   ScaleS = sqrt(2);
   Family = 'HSV design';
case {'s+p(4,2)','lc9/7-m'}
```

```
    Seq = {0,0;[1,-9,-9,1]/16,2;[1,1]/4,0};
    ScaleS = sqrt(2);
    Family = 'S+P';
case 's+p(6,2)'
    Seq = {0,0;[-3,25,-150,-150,25,-
    3]/256,3;[1,1]/4,0};
    ScaleS = sqrt(2);
    Family = 'S+P';
case {'s+p(4,4)','lc13/7-t'}
    Seq = {0,0;[1,-9,-9,1]/16,2;[-1,9,9,-1]/32,1};
    ScaleS = sqrt(2);
    Family = 'S+P';
case {'s+p(2+2,2)','lc5/11-c'}
    Seq = {0,0;[-1,-1]/2,1;[1,1]/4,0;-[-1,1,1,-
    1]/16,2};
    ScaleS = sqrt(2);
    Family = 'S+P';
case 'tt'
    Seq = {1,0;[3,-22,-128,22,-3]/256,2};
    ScaleD = sqrt(2);
    Family = 'Le Gall-Tabatabai polynomial';
case 'lc2/6'
    Seq = {0,0;-1,0;1/2,0;[-1,0,1]/4,1};
    ScaleS = sqrt(2);
    ScaleD = -1/sqrt(2);
    Family = 'Reverse spline';
case 'lc2/10'
    Seq = {0,0;-1,0;1/2,0;[3,-22,0,22,-3]/64,2};
    ScaleS = sqrt(2);
    ScaleD = -1/sqrt(2);
    Family = 'Reverse spline';
case 'lc5/11-a'
    Seq = {0,0;-[1,1]/2,1;[1,1]/4,0;[1,-1,-1,1]/32,2};
    ScaleS = sqrt(2);
    ScaleD = -1/sqrt(2);
    Family - 'Low complcxity';
case 'lc6/14'
    Seq = {0,0;-1,0;[-1,8,1]/16,1;[1,-6,0,6,-1]/16,2};
    ScaleS = sqrt(2);
    ScaleD = -1/sqrt(2);
    Family = 'Low complexity';
case 'lc13/7-c'
    Seq = {0,0;[1,-9,-9,1]/16,2;[-1,5,5,-1]/16,1};
    ScaleS = sqrt(2);
    ScaleD = -1/sqrt(2);
    Family = 'Low complexity';
otherwise
    Seq = {};
    return;
end
```

```
if ~isempty(findstr(lower(WaveletName),'rspline'))
   [Seq,ScaleS,ScaleD] = seqdual(Seq,ScaleS,ScaleD);
   Family = 'Reverse spline';
end

return;


function [Seq,ScaleS,ScaleD] = seqdual(Seq,ScaleS,ScaleD)
% Dual of a lifting sequence

L = size(Seq,1);

for k = 1:L
   % f'(z) = -f(z^-1)
   Seq{k,2} = -(Seq{k,2} - length(Seq{k,1}) + 1);
   Seq{k,1} = -fliplr(Seq{k,1});
end

if all(Seq{1,1} == 0)
   Seq = reshape({Seq{2:end,:}},L-1,2);
else
   [Seq{1:L+1,:}] = deal(0,Seq{1:L,1},0,Seq{1:L,2});
end

ScaleS = 1/ScaleS;
ScaleD = 1/ScaleD;
return;


function [h,g] = seq2hg(Seq,ScaleS,ScaleD,Dual)
% Find wavelet filters from lifting sequence
if Dual, [Seq,ScaleS,ScaleD] =
seqdual(Seq,ScaleS,ScaleD); end
if rem(size(Seq,1),2), [Seq{size(Seq,1)+1,:}] =
deal(0,0); end

h = {1,0};
g = {1,1};

for k = 1:2:size(Seq,1)
   h = lp_lift(h,g,{Seq{k,:}});
   g = lp_lift(g,h,{Seq{k+1,:}});
end

h = {ScaleS*h{1},h{2}};
g = {ScaleD*g{1},g{2}};

if Dual
   h{2} = -(h{2} - length(h{1}) + 1);
   h{1} = fliplr(h{1});
```

```
    g{2} = -(g{2} - length(g{1}) + 1);
    g{1} = fliplr(g{1});
end

return;


function a = lp_lift(a,b,c)
% a(z) = a(z) + b(z) c(z^2)

d = zeros(1,length(c{1})*2-1);
d(1:2:end) = c{1};
d = conv(b{1},d);
z = b{2}+c{2}*2;
zmax = max(a{2},z);
f = [zeros(1,zmax-a{2}),a{1},zeros(1,a{2} -
length(a{1}) - z + length(d))];
i = zmax-z + (1:length(d));
f(i) = f(i) + d;

if all(abs(f) < 1e-12)
    a = {0,0};
else
    i = find(abs(f)/max(abs(f)) > 1e-10);
    i1 = min(i);
    a = {f(i1:max(i)),zmax-i1+1};
end
return;


function X = xfir(B,Z,X,Ext)
%XFIR  Noncausal FIR filtering with boundary handling.
%   Y = XFIR(B,Z,X,EXT) filters X with FIR filter B with
    leading
%   delay -Z along the columns of X.  EXT specifies the
    boundary
%   handling.  Special handling  is done for one and
    two-tap filters.

% Pascal Getreuer 2005-2006

N = size(X);

% Special handling for short filters
if length(B) == 1 & Z == 0
    if B == 0
        X = zeros(size(X));
    elseif B ~= 1
        X = B*X;
    end
    return;
end
```

```
% Compute the number of samples to add to each end of
  the signal
pl = max(length(B)-1-Z,0);  % Padding on the left end
pr = max(Z,0);              % Padding on the right end

switch lower(Ext)
case {'sym','wsws'}   % Symmetric extension, WSWS
   if all([pl,pr] < N(1))
         X = filter(B,1,X([pl+1:-1:2,1:N(1),N(1)-1:-
         1:N(1)-pr],:,:),[],1);
         X = X(Z+pl+1:Z+pl+N(1),:,:);
      return;
   else
      i = [1:N(1),N(1)-1:-1:2];
      Ns = 2*N(1) - 2 + (N(1) == 1);
      i = i([rem(pl*(Ns-1):pl*Ns-
      1,Ns)+1,1:N(1),rem(N(1):N(1)+pr-1,Ns)+1]);
   end
case {'symh','hshs'}  % Symmetric extension, HSHS
   if all([pl,pr] < N(1))
      i = [pl:-1:1,1:N(1),N(1):-1:N(1)-pr+1];
   else
      i = [1:N(1),N(1):-1:1];
      Ns = 2*N(1);
      i = i([rem(pl*(Ns-1):pl*Ns-
      1,Ns)+1,1:N(1),rem(N(1):N(1)+pr-1,Ns)+1]);
   end
case 'wshs'           % Symmetric extension, WSHS
   if all([pl,pr] < N(1))
      i = [pl+1:-1:2,1:N(1),N(1):-1:N(1)-pr+1];
   else
      i = [1:N(1),N(1):-1:2];
      Ns = 2*N(1) - 1;
      i = i([rem(pl*(Ns-1):pl*Ns-
      1,Ns)+1,1:N(1),rem(N(1):N(1)+pr-1,Ns)+1]);
   end
case 'hsws'           % Symmetric extension, HSWS
   if all([pl,pr] < N(1))
      i = [pl:-1:1,1:N(1),N(1)-1:-1:N(1)-pr];
   else
      i = [1:N(1),N(1)-1:-1:1];
      Ns = 2*N(1) - 1;
      i = i([rem(pl*(Ns-1):pl*Ns-
      1,Ns)+1,1:N(1),rem(N(1):N(1)+pr-1,Ns)+1]);
   end
case 'zpd'
   Ml = N; Ml(1) = pl;
   Mr = N; Mr(1) = pr;

   X = filter(B,1,[zeros(Ml);X;zeros(Mr)],[],1);
```

```
   X = X(Z+pl+1:Z+pl+N(1),:,:);
   return;
case 'per'               % Periodic
   i = [rem(pl*(N(1)-1):pl*N(1)-
   1,N(1))+1,1:N(1),rem(0:pr-1,N(1))+1];
case 'sp0'               % Constant extrapolation
   i = [ones(1,pl),1:N(1),N(1)+zeros(1,pr)];
   case 'asym'           % Asymmetric extension
   i1 = [ones(1,pl),1:N(1),N(1)+zeros(1,pr)];

   if all([pl,pr] < N(1))
      i2 = [pl+1:-1:2,1:N(1),N(1)-1:-1:N(1)-pr];
   else
      i2 = [1:N(1),N(1)-1:-1:2];
      Ns = 2*N(1) - 2 + (N(1) == 1);
      i2 = i2([rem(pl*(Ns-1):pl*Ns-
      1,Ns)+1,1:N(1),rem(N(1):N(1)+pr-1,Ns)+1]);
   end

   X = filter(B,1,2*X(i1,:,:) - X(i2,:,:),[],1);
   X = X(Z+pl+1:Z+pl+N(1),:,:);
   return;
otherwise
   error(['Unknown boundary handling, ''',Ext,'''.']);
end

X = filter(B,1,X(i,:,:),[],1);
X = X(Z+pl+1:Z+pl+N(1),:,:);
return;


function [x1,phi,x2,psi] = cascade(h,g,dx)
% Wavelet cascade algorithm

c = h{1}*2/sum(h{1});
x = 0:dx:length(c) - 1;
x1 = x - h{2};
phi0 = 1 - abs(linspace(-1,1,length(x))).';

ii = []; jj = []; s = [];

for k = 1:length(c)
   xk = 2*x - (k-1);
   i = find(xk >= 0 & xk <= length(c) - 1);
   ii = [ii,i];
   jj = [jj,floor(xk(i)/dx)+1];
   s = [s,c(k)+zeros(size(i))];
end

% Construct a sparse linear operator that iterates the
  dilation equation
Dilation = sparse(ii,jj,s,length(x),length(x));
```

```
for N = 1:30
   phi = Dilation*phi0;
   if norm(phi - phi0,inf) < 1e-5, break; end
   phi0 = phi;
end

if norm(phi) == 0
   phi = ones(size(phi))*sqrt(2);    % Special case for
Haar scaling function
else
   phi = phi/(norm(phi)*sqrt(dx));  % Rescale result
end

if nargout > 2
   phi2 = phi(1:2:end);   % phi2 is approximately
phi(2x)

   if length(c) == 2
      L = length(phi2);
   else
      L = ceil(0.5/dx);
   end

   % Construct psi from translates of phi2
   c = g{1};
   psi = zeros(length(phi2)+L*(length(c)-1),1);
   x2 = (0:length(psi)-1)*dx - g{2} - 0*h{2}/2;

   for k = 1:length(c)
      i = (1:length(phi2)) + L*(k-1);
      psi(i) = psi(i) + c(k)*phi2;
   end
end
return;


function s = filterstr(a,K)
% Convert a filter to a string

[n,d] = rat(K/sqrt(2));

if d < 50
   a{1} = a{1}/sqrt(2);    % Scale filter by sqrt(2)
   s = '( ';
else
   s = '';
end

Scale = [pow2(1:15),10,20,160,280,inf];

for i = 1:length(Scale)
   if norm(round(a{1}*Scale(i))/Scale(i) - a{1},inf)
   < 1e-9
```

```
        a{1} = a{1}*Scale(i);  % Scale filter by a power
        of 2 or 160
        s = '( ';
        break;
    end
end

z = a{2};
LineOff = 0;

for k = 1:length(a{1})
    v = a{1}(k);
    if v ~= 0  % Only display nonzero coefficients
        if k > 1
            s2 = [' ',char(44-sign(v)),' '];
            v = abs(v);
        else
            s2 = '';
        end

        s2 = sprintf('%s%g',s2,v);

        if z == 1
            s2 = sprintf('%s z',s2);
        elseif z ~= 0
            s2 = sprintf('%s z^%d',s2,z);
        end

        if length(s) + length(s2) > 72 + LineOff  % Wrap
        long lines
            s2 = [char(10),'          ',s2];
            LineOff = length(s);
        end

        s = [s,s2];
    end

    z = z - 1;
end

if s(1) == '('
    s = [s,' )'];
    if d < 50, s = [s,' sqrt(2)']; end
    if i < length(Scale)
        s = sprintf('%s/%d',s,Scale(i));
    end
end

return;


function N = numvanish(g)
% Determine the number of vanishing moments from
highpass filter g(z)
```

```
for N = 0:length(g)-1  % Count the number of roots at
z = 1
   [g,r] = deconv(g,[1,-1]);
   if norm(r,inf) > 1e-7, break; end
end
return;

function [iout,qout]=delay(idata,qdata,nsamp,idel)
iout=zeros(1,nsamp);
qout=zeros(1,nsamp);
if idel ~= 0
    iout(1:idel)=zeros(1,idel);
    qout(1:idel)=zeros(1,idel);
end
iout(idel+1:nsamp)=idata(1:nsamp-idel);
qout(idel+1:nsamp)=qdata(1:nsamp-idel);

function  [iout,qout]=comb(idata,qdata,attn)
iout=randn(1,length(idata)).*attn;
qout=randn(1,length(qdata)).*attn;
iout=iout+idata(1:length(idata));
qout=qout+qdata(1:length(qdata));

function [iout,qout]=giins(idata,qdata,fftlen,
gilen,nd);
idata1=reshape(idata,fftlen,nd);
qdata1=reshape(qdata,fftlen,nd);
idata2=[idata1(fftlen-gilen+1:fftlen,:);idata1];
qdata2=[qdata1(fftlen-gilen+1:fftlen,:);qdata1];
iout=reshape(idata2,1,(fftlen+gilen)*nd);
qout=reshape(qdata2,1,(fftlen+gilen)*nd);

function [iout,qout]=girem(idata,qdata,fftlen2,
gilen,nd);
idata2=reshape(idata,fftlen2,nd);
qdata2=reshape(qdata,fftlen2,nd);
iout=idata2(gilen+1:fftlen2,:);
qout=qdata2(gilen+1:fftlen2,:);

function [iout,qout]=giins1(idata,qdata,fftlen,
gilen,nd);
idata1=reshape(idata,fftlen,nd);
qdata1=reshape(qdata,fftlen,nd);
gg=gilen/2;
idata2=[zeros(gg,nd);idata1(fftlen-
gg+1:fftlen,:);idata1];
qdata2=[zeros(gg,nd);qdata1(fftlen-
gg+1:fftlen,:);qdata1];
```

```
iout=reshape(idata2,1,(fftlen+gilen)*nd);
qout=reshape(qdata2,1,(fftlen+gilen)*nd);

function[iout,qout,ramp,rcos,rsin]=sefade(idata,qdata,
itau,dlvl,th,n0,itn,n1,nsamp,tstp,fd,flat)
iout=zeros(1,nsamp);;
qout=zeros(1,nsamp);;
total_attn=sum(10.^(-1.0.*dlvl./10.0));
for k=1:n1
    atts=10.^(-0.05.*dlvl(k));
    theta=th(k).*pi./180.0;
    [itmp,qtmp]=delay(idata,qdata,nsamp,itau(k));
     [itmp3,qtmp3,ramp,rcos,rsin]=fade(itmp,qtmp,nsamp
    ,tstp,fd,n0(k),itn(k),flat);
    iout=iout+atts.*itmp3./sqrt(total_attn);
    qout=qout+atts.*qtmp3./sqrt(total_attn);
end

function [iout,qout]=girem1(idata,qdata,fftlen2,
gilen,nd);
idata2=reshape(idata,fftlen2,nd);
qdata2=reshape(qdata,fftlen2,nd);
iout=idata2(gilen+1:fftlen2,:);
qout=qdata2(gilen+1:fftlen2,:);

function [iout,qout]=giins1(idata,qdata,fftlen,
gilen,nd);
idata1=reshape(idata,fftlen,nd);
qdata1=reshape(qdata,fftlen,nd);
idata2=[zeros(gilen,nd);idata1];
qdata2=[zeros(gilen,nd);qdata1];
iout=reshape(idata2,1,(fftlen+gilen)*nd);
qout=reshape(qdata2,1,(fftlen+gilen)*nd);
```

Appendix C: MATLAB® Codes for Image Encryption

```
function y=multi(x,w)
%z=a*a1 mod 2^32
a=frombase256(x);
a1=frombase256(w);
z=mod(a*a1,2^32);
y=tobase256(z);

function permutation=p(substitution)
% p function permutes an input "substitution" based on x
x=[16 7 20 21 29 12 28 17 01 15 23 26 05 18 31 10 2 8
24 14 32 27 3 9 19 13 30 6 22 11 4 25];
permutation=substitution(x);

function y=padding(x)
%Padding x with zeros if x is shorter than 16
L=length(x);
if mod(L,16)==0
  y=x;
else
  pad=16-mod(L,16);
  y=zeros(1,L+pad);
  y(1:L)=x;
end;

function RC5enctextRC5=RC5CBC(plaintext1,m,n,CO,r,keyi)
% Encrypt data using RC5 in CBC mode.
RC5enctextRC5=zeros(m*n/8,8);
for i=1:m*n/8
  plaintext=plaintext1(i,:);
  if i==1
    plaintexts=bitxor(plaintext,CO);
  else
    plaintexts=bitxor(plaintext,RC5enctext);
  end;
  RC5enctext = RC5enc(plaintexts,r,keyi);
  RC5enctextRC5(i,:)=RC5enctext;
end;
```

```
function plaintextRC5=RC5CBCDec(RC5enctext1,
m,n,CO,r,keyi)
% Decrypt data using RC5 in CBC mode.
plaintextRC5=zeros(m*n/8,8);
for i=1:m*n/8
  RC5enctextss=RC5enctext1(i,:);
  re_plaintext = RC5decry(RC5enctextss,r,keyi);
  if i==1
    plaintext=bitxor(re_plaintext,CO);
  else
    plaintext=bitxor(re_plaintext,RC5enctext1((i-1),:));
  end;
plaintextRC5(i,:)=plaintext;
end;

function RC5enctextRC5=RC5CFB(plaintext1,m,n,CO,r,keyi)
% Encrypt data using RC5 in CFB mode.
RC5enctextRC5=zeros(m*n/8,8);
for i=1:m*n/8
  plaintext=plaintext1(i,:);
  if i==1
    cr1=RC5enc(CO,r,keyi);
    RC5enctext=bitxor(plaintext,cr1);
  else
    cr1=RC5enc(RC5enctext,r,keyi);
    RC5enctext=bitxor(plaintext,cr1);
  end;
  RC5enctextRC5(i,:)=RC5enctext;
end;

function plaintextRC5=RC5CFBDec(RC5enctext1,
m,n,CO,r,keyi)
% Decrypt data using RC5 in CFB mode.
plaintextRC5=zeros(m*n/8,8);
for i=1:m*n/8
  RC5enctextss=RC5enctext1(i,:);
  if i==1
    cr1=RC5enc(CO,r,keyi);
    re_plaintext=bitxor(RC5enctextss,cr1);
  else
    cr1=RC5enc(RC5enctext1(i-1,:),r,keyi);
    re_plaintext=bitxor(RC5enctextss,cr1);
  end;
  plaintextRC5(i,:)=re_plaintext;
  end;

function [xa1, Fs, nbits]=RC5DecCBCAudio
(x1, Fs, nbits,key,r,CO)
% Decrypt Audio using RC5 in CBC mode.
```

```
keyi=RC5keygen(key,r);
%CBC L=length(x1);
y1=double(x1);
plaintextRC5=zeros(L/8,8);
RC5enctext1=reshape(y1,L/8,8);
for i=1:L/8
  RC5enctextss=RC5enctext1(i,:);
  re_plaintext = RC5decry(RC5enctextss,r,keyi);
  if i==1
    plaintext=bitxor(re_plaintext,CO);
  else
    plaintext=bitxor(re_plaintext,RC5enctext1((i-1),:));
  end;
  plaintextRC5(i,:)=plaintext;
end;
xa1=uint8(plaintextRC5(:));

function xa1=RC5DecCBCImage(x1,key,r,CO)
% Decrypt Image using RC5 in CBC mode.
keyi=RC5keygen(key,r);
%CBC
[m,n]=size(x1);
y1=double(x1);
plaintextRC5=zeros(m*n/8,8);
RC5enctext1=reshape(y1',8,m*n/8)';
for i=1:m*n/8
  RC5enctextss=RC5enctext1(i,:);
  re_plaintext = RC5decry(RC5enctextss,r,keyi);
  if i==1
    plaintext=bitxor(re_plaintext,CO);
  else
    plaintext=bitxor(re_plaintext,RC5enctext1((i-1),:));
  end;
  plaintextRC5(i,:)=plaintext;
end;
ya1=reshape(plaintextRC5',n,m)';
xa1=uint8(ya1);

function xa1=RC5DecCBCimageC(x,key,r,CO)
% Decrypt Colored Image using RC5 in CBC mode.
keyi=RC5keygen(key,r);
%CBC x1=x(:,:,1);
x2=x(:,:,2);
x3=x(:,:,3);
y1=double(x1);
y2=double(x2);
y3=double(x3);
[m,n]=size(x1);
plaintext(:,:,1)=reshape(y1',8,m*n/8)';
```

```
plaintext(:,:,2)=reshape(y2',8,m*n/8)';
plaintext(:,:,3)=reshape(y3',8,m*n/8)';
ciphertext=x;
for i=1:3 plaintextRC5=plaintext(:,:,i);
  RC5enctext=RC5CBCDec(plaintextRC5,m,n,CO,r,keyi);
  RC5enctext=reshape(RC5enctext',n,m)';
  ciphertext(:,:,i)=RC5enctext;
end
xa1=uint8(ciphertext);

function [xa1, Fs, nbits]=RC5DecCFBAudio(x1, Fs,
nbits,key,r,CO)
%Decrypt Audio using RC5 in CFB mode.
keyi=RC5keygen(key,r);
L=length(x1); y1=double(x1);
plaintextRC5=zeros(L/8,8);
RC5enctext1=reshape(y1,L/8,8);
for i=1:L/8
  RC5enctextss=RC5enctext1(i,:);
  if i==1
    cr1=RC5enc(CO,r,keyi);
    re_plaintext=bitxor(RC5enctextss,cr1);
  else
    cr1=RC5enc(RC5enctext1(i-1,:),r,keyi);
    re_plaintext=bitxor(RC5enctextss,cr1);
  end;
  plaintextRC5(i,:)=re_plaintext;
end;
xa1=uint8(plaintextRC5(:));

function xa1=RC5DecCFBImage(x1,key,r,CO)
%Decrypt Image using RC5 in CFB mode.
keyi=RC5keygen(key,r);
[m,n]=size(x1); y1=double(x1);
plaintextRC5=zeros(m*n/8,8);
RC5enctext1=reshape(y1',8,m*n/8)';
for i=1:m*n/8
  RC5enctextss=RC5enctext1(i,:);
  if i==1
    cr1=RC5enc(CO,r,keyi);
    re_plaintext=bitxor(RC5enctextss,cr1);
  else
    cr1=RC5enc(RC5enctext1(i-1,:),r,keyi);
    re_plaintext=bitxor(RC5enctextss,cr1);
  end;
    plaintextRC5(i,:)=re_plaintext;
end;
ya1=reshape(plaintextRC5',n,m)';
xa1=uint8(ya1);
```

```
function xa1=RC5DecCFBImageC(x,key,r,CO)
%Decrypt Colored Image using RC5 in CFB mode.
keyi=RC5keygen(key,r);
%CFB
x1=x(:,:,1);
x2=x(:,:,2);
x3=x(:,:,3);
y1=double(x1);
y2=double(x2);
y3=double(x3);
[m,n]=size(x1);
plaintext(:,:,1)=reshape(y1',8,m*n/8)';
plaintext(:,:,2)=reshape(y2',8,m*n/8)';
plaintext(:,:,3)=reshape(y3',8,m*n/8)';
ciphertext=x;
for i=1:3 plaintextRC5=plaintext(:,:,i);
  RC5enctext=RC5CFBDec(plaintextRC5,m,n,CO,r,keyi);
  RC5enctext=reshape(RC5enctext',n,m)';
  ciphertext(:,:,i)=RC5enctext;
end
xa1=uint8(ciphertext);

function [xa1, Fs, nbits]=RC5DecECBAudio(x1, Fs,
nbits,key,r)
%Decrypt Audio using RC5 in ECB mode.
keyi=RC5keygen(key,r);
%ECB L=length(x1);
y1=double(x1);
plaintextRC5=zeros(L/8,8);
RC5enctext1=reshape(y1,L/8,8);
for i=1:L/8
  RC5enctextss=RC5enctext1(i,:);
  re_plaintext =RC5decry(RC5enctextss,r,keyi);
  plaintextRC5(i,:)=re_plaintext;
end;
xa1=uint8(plaintextRC5(:));

function xa1=RC5DecECBImage(x1,key,r)
%Decrypt Image using RC5 in ECB mode.
keyi=RC5keygen(key,r);
[m,n]=size(x1); y1=double(x1);
plaintextRC5=zeros(m*n/8,8);
RC5enctext1=reshape(y1',8,m*n/8)';
for i=1:m*n/8
  RC5enctextss=RC5enctext1(i,:);
  re_plaintext =RC5decry(RC5enctextss,r,keyi);
  plaintextRC5(i,:)=re_plaintext;
end;
```

```
ya1=reshape(plaintextRC5',n,m)';
xa1=uint8(ya1);

function xa1=RC5DecECBImageC(x,key,r)
%Decrypt Colored Image using RC5 in ECB mode.
keyi=RC5keygen(key,r);
%ECB
x1=x(:,:,1);
x2=x(:,:,2);
x3=x(:,:,3);
y1=double(x1);
y2=double(x2);
y3=double(x3);
[m,n]=size(x1);
plaintext(:,:,1)=reshape(y1',8,m*n/8)';
plaintext(:,:,2)=reshape(y2',8,m*n/8)';
plaintext(:,:,3)=reshape(y3',8,m*n/8)';
ciphertext=x;
for i=1:3 plaintextRC5=plaintext(:,:,i);
  RC5enctext=RC5ECBDec(plaintextRC5,m,n,r,keyi);
  RC5enctext=reshape(RC5enctext',n,m)';
  ciphertext(:,:,i)=RC5enctext;
end
xa1=uint8(ciphertext);

function [xa1, Fs, nbits]=RC5DecOFBAudio(x1, Fs,
nbits,key,r,CO)
%Decrypt Audio using RC5 in OFB mode.
keyi=RC5keygen(key,r);
%OFB
L=length(x1);
y1=double(x1);
plaintextRC5=zeros(L/8,8);
RC5enctext1=reshape(y1,L/8,8);
for i=1:L/8
  RC5enctextss=RC5enctext1(i,:);
  if i==1
    cr=RC5enc (CO,r,keyi);
    re_plaintext=bitxor(RC5enctextss,cr);
  else cr=RC5enc(cr,r,keyi);
    re_plaintext=bitxor(RC5enctextss,cr);
  end;
  plaintextRC5(i,:)=re_plaintext;
end;
xa1=uint8(plaintextRC5(:));

function xa1=RC5DecOFBImage(x1,key,r,CO)
%Decrypt Image using RC5 in OFB mode.
keyi=RC5keygen(key,r);
```

```
[m,n]=size(x1); y1=double(x1);
plaintextRC5=zeros(m*n/8,8);
RC5enctext1=reshape(y1',8,m*n/8)';
for i=1:m*n/8
  RC5enctextss=RC5enctext1(i,:);
  if i==1
    cr=RC5enc (CO,r,keyi);
    re_plaintext=bitxor(RC5enctextss,cr);
  else cr=RC5enc(cr,r,keyi);
    re_plaintext=bitxor(RC5enctextss,cr);
  end;
  plaintextRC5(i,:)=re_plaintext;
end;
ya1=reshape(plaintextRC5',n,m)';
xa1=uint8(ya1);

function xa1=RC5DecOFBImagec(x,key,r,CO)
%Decrypt Colored Image using RC5 in OFB mode.
keyi=RC5keygen(key,r);
%OFB
x1=x(:,:,1);
x2=x(:,:,2);
x3=x(:,:,3);
y1=double(x1);
y2=double(x2);
y3=double(x3);
[m,n]=size(x1);
plaintext(:,:,1)=reshape(y1',8,m*n/8)';
plaintext(:,:,2)=reshape(y2',8,m*n/8)';
plaintext(:,:,3)=reshape(y3',8,m*n/8)';
ciphertext=x;
for i=1:3 plaintextRC5=plaintext(:,:,i);
  RC5enctext=RC5OFBDec(plaintextRC5,m,n,CO,r,keyi);
  RC5enctext=reshape(RC5enctext',n,m)';
  ciphertext(:,:,i)=RC5enctext;
end
xa1=uint8(ciphertext);

function y=RC5decry(plaintext,round,s)
%RC5 Decryption
  a=plaintext(1:4); b=plaintext(5:8);
  for i=round:-1:1
    b=RC5sub(b,s(2*i+2,:)); b=shifting(b,-LSB5(a));
    b=bitxor(b',a); a=RC5sub(a,s(2*i+1,:));
    a=shifting(a,-LSB5(b)); a=bitxor(a',b);
  end b=RC5sub(b,s(2,:));
  a=RC5sub(a,s(1,:)); y(1:4)=a;
  y(5:8)=b;
```

```
function RC5enctextRC5=RC5ECB(plaintext1,m,n,r,keyi)
%Encrypt data using RC5 in ECB mode
RC5enctextRC5=zeros(m*n/8,8);
for i=1:m*n/8 plaintext=plaintext1(i,:);
  RC5enctext = RC5enc(plaintext,r,keyi);
  RC5enctextRC5(i,:)=RC5enctext;
end;

function plaintextRC5=RC5ECBDec(RC5enctext1,m,n,r,keyi)
%Decrypt data using RC5 in ECB mode
plaintextRC5=zeros(m*n/8,8);
for i=1:m*n/8
  RC5enctextss=RC5enctext1(i,:);
  re_plaintext =RC5decry(RC5enctextss,r,keyi);
  plaintextRC5(i,:)=re_plaintext;
end;

function y=RC5enc(plaintext,round,s)
%RC5 Encryption
a=plaintext(1:4); b=plaintext(5:8);
a=add(a,s(1,:)); b=add(b,s(2,:)); for
i=1:round
  a=bitxor(a,b); a=shifting(a,LSB5(b));
  a=add(a,s(2*i+1,:)); b=bitxor(b,a);
  b=shifting(b,LSB5(a)); b=add(b,s(2*i+2,:));
end y(1:4)=a; y(5:8)=b;

function [xa, Fs, nbits]=RC5EncCBCAudio(x, Fs,
nbits,key,r,CO)
%Encrypt Audio using RC5 in CBC mode
L=length(x);
y=double(x); keyi=RC5keygen(key,r);
plaintext1=reshape(y,L/8,8);
RC5enctextRC5=zeros(L/8,8);
  %CBC
for i=1:L/8 plaintext=plaintext1(i,:); if i==1
    plaintexts=bitxor(plaintext,CO);
  else plaintexts=bitxor(plaintext,RC5enctext);
  end;
  RC5enctext = RC5enc(plaintexts,r,keyi);
  RC5enctextRC5(i,:)=RC5enctext;
end;
xa=uint8(RC5enctextRC5(:));

function xa=RC5EncCBCImage(x,key,r,CO)
%Encrypt Image using RC5 in CBC mode
[m,n]=size(x);
y=double(x); keyi=RC5keygen(key,r);
plaintext1=reshape(y',8,m*n/8)';
```

```
RC5enctextRC5=zeros(m*n/8,8);
  %CBC
for i=1:m*n/8 plaintext=plaintext1(i,:); if i==1
    plaintexts=bitxor(plaintext,CO);
  else plaintexts=bitxor(plaintext,RC5enctext);
  end;
  RC5enctext = RC5enc(plaintexts,r,keyi);
  RC5enctextRC5(i,:)=RC5enctext;
end; ya=reshape(RC5enctextRC5',n,m)'; xa=uint8(ya);

function xa=RC5EncCBCImageC(x,key,r,CO)
%Encrypt Colored Image using RC5 in CBC mode
x1=x(:,:,1); x2=x(:,:,2); x3=x(:,:,3);
y1=double(x1); y2=double(x2);
y3=double(x3); [m,n]=size(x1);
keyi=RC5keygen(key,r);
plaintext(:,:,1)=reshape(y1',8,m*n/8)';
plaintext(:,:,2)=reshape(y2',8,m*n/8)';
plaintext(:,:,3)=reshape(y3',8,m*n/8)';
RC5enctext=x;
%CBC
for i=1:3 plaintext1=plaintext(:,:,i);
  RC5enctextRC5=RC5CBC(plaintext1,m,n,CO,r,keyi);
  RC5enctextRC5=reshape(RC5enctextRC5',n,m)';
  RC5enctext(:,:,i)=RC5enctextRC5;
end;
xa=uint8(RC5enctext);

function [xa, Fs, nbits]=RC5EncCFBAudio(x, Fs,
nbits,key,r,CO)
%Encrypt Audio using RC5 in CBC mode
L=length(x);
y=double(x); keyi=RC5keygen(key,r);
plaintext1=reshape(y,L/8,8);
RC5enctextRC5=zeros(L/8,8);
 % CFB
for i=1:L/8 plaintext=plaintext1(i,:); if i==1
    cr1=RC5enc(CO,r,keyi);
    RC5enctext=bitxor(plaintext,cr1);
  else cr1=RC5enc(RC5enctext,r,keyi);
    RC5enctext=bitxor(plaintext,cr1);
  end;
    RC5enctextRC5(i,:)=RC5enctext;
end;
xa=uint8(RC5enctextRC5(:));

function xa=RC5EncCFBImage(x,key,r,CO)
%Encrypt Image using RC5 in CFB mode
```

```
[m,n]=size(x);
y=double(x); keyi=RC5keygen(key,r);
plaintext1=reshape(y',8,m*n/8)';
RC5enctextRC5=zeros(m*n/8,8);
for i=1:m*n/8 plaintext=plaintext1(i,:); if i==1
    cr1=RC5enc(CO,r,keyi);
    RC5enctext=bitxor(plaintext,cr1);
  else cr1=RC5enc(RC5enctext,r,keyi);
    RC5enctext=bitxor(plaintext,cr1);
  end;
  RC5enctextRC5(i,:)=RC5enctext;
end; ya=reshape(RC5enctextRC5',n,m)'; xa=uint8(ya);

function xa=RC5EncCFBImageC(x,key,r,CO)
%Encrypt Colored Image using RC5 in CFB mode
x1=x(:,:,1); x2=x(:,:,2); x3=x(:,:,3);
y1=double(x1); y2=double(x2);
y3=double(x3); [m,n]=size(x1);
keyi=RC5keygen(key,r);
plaintext(:,:,1)=reshape(y1',8,m*n/8)';
plaintext(:,:,2)=reshape(y2',8,m*n/8)';
plaintext(:,:,3)=reshape(y3',8,m*n/8)';
RC5enctext=x;
% CFB
for i=1:3 plaintext1=plaintext(:,:,i);
  RC5enctextRC5=RC5CFB(plaintext1,m,n,CO,r,keyi);
  RC5enctextRC5=reshape(RC5enctextRC5',n,m)';
  RC5enctext(:,:,i)=RC5enctextRC5;
end; xa=uint8(RC5enctext);
imwrite(xa,'onion1RC5encCFB.tif','tif')
imshow('onion1RC5encCFB.tif')

function [xa, Fs, nbits]=RC5EncECBAudio(x, Fs,
nbits,key,r)
%Encrypt Audio using RC5 in ECB mode
L=length(x);
y=double(x); keyi=RC5keygen(key,r);
plaintext1=reshape(y,L/8,8);
RC5enctextRC5=zeros(L/8,8);
% ECB
for i=1:L/8 plaintext=plaintext1(i,:);
  RC5enctext = RC5enc(plaintext,r,keyi);
  RC5enctextRC5(i,:)=RC5enctext;
end;
xa=uint8(RC5enctextRC5(:));

function xa=RC5EncECBImage(x,key,r)
%Encrypt Image using RC5 in ECB mode
```

```
[m,n]=size(x);
y=double(x); keyi=RC5keygen(key,r);
plaintext1=reshape(y',8,m*n/8)';
RC5enctextRC5=zeros(m*n/8,8);
 % ECB
for i=1:m*n/8
plaintext=plaintext1(i,:);
  RC5enctext = RC5enc(plaintext,r,keyi);
  RC5enctextRC5(i,:)=RC5enctext;
end; ya=reshape(RC5enctextRC5',n,m)'; xa=uint8(ya);

function xa=RC5EncECBImageC(x,key,r)
%Encrypt Colored Image using RC5 in ECB mode
x1=x(:,:,1); x2=x(:,:,2); x3=x(:,:,3);
y1=double(x1); y2=double(x2);
y3=double(x3); [m,n]=size(x1);
keyi=RC5keygen(key,r);
plaintext(:,:,1)=reshape(y1',8,m*n/8)';
plaintext(:,:,2)=reshape(y2',8,m*n/8)';
plaintext(:,:,3)=reshape(y3',8,m*n/8)';
RC5enctext=x;% ECB
for i=1:3 plaintext1=plaintext(:,:,i);
  RC5enctextRC5=RC5ECB(plaintext1,m,n,r,keyi);
  RC5enctextRC5=reshape(RC5enctextRC5',n,m)';
  RC5enctext(:,:,i)=RC5enctextRC5;
end;
xa=uint8(RC5enctext);

function [xa, Fs, nbits]=RC5EncOFBAudio(x, Fs,
nbits,key,r,CO)
%Encrypt Audio using RC5 in OFB mode
L=length(x);
y=double(x); keyi=RC5keygen(key,r);
plaintext1=reshape(y,L/8,8);
RC5enctextRC5=zeros(L/8,8);
% OFB
for i=1:L/8 plaintext=plaintext1(i,:);
  if i==1 cr=RC5enc(CO,r,keyi);
  else cr=RC5enc(cr,r,keyi);
  end; RC5enctext=bitxor(plaintext,cr);
  RC5enctextRC5(i,:)=RC5enctext;
end;
xa=uint8(RC5enctextRC5(:));

function xa=RC5EncOFBImage(x,key,r,CO)
%Encrypt Image using RC5 in OFB mode
 [m,n]=size(x);
y=double(x); keyi=RC5keygen(key,r);
```

```
plaintext1=reshape(y',8,m*n/8)';
RC5enctextRC5=zeros(m*n/8,8);
 % OFB
for i=1:m*n/8 plaintext=plaintext1(i,:);
  if i==1 cr=RC5enc(CO,r,keyi);
  else cr=RC5enc(cr,r,keyi);
  end; RC5enctext=bitxor(plaintext,cr);
  RC5enctextRC5(i,:)=RC5enctext;
end; ya=reshape(RC5enctextRC5',n,m)'; xa=uint8(ya);

function xa=RC5EncOFBImageC(x,key,r,CO)
%Encrypt Colored Image using RC5 in OFB mode
x1=x(:,:,1); x2=x(:,:,2); x3=x(:,:,3);
y1=double(x1); y2=double(x2);
y3=double(x3); [m,n]=size(x1);
keyi=RC5keygen(key,r);
plaintext(:,:,1)=reshape(y1',8,m*n/8)';
plaintext(:,:,2)=reshape(y2',8,m*n/8)';
plaintext(:,:,3)=reshape(y3',8,m*n/8)'; RC5enctext=x;
% OFB
for i=1:3 plaintext1=plaintext(:,:,i);
  RC5enctextRC5=RC5OFB(plaintext1,m,n,CO,r,keyi);
  RC5enctextRC5=reshape(RC5enctextRC5',n,m)';
  RC5enctext(:,:,i)=RC5enctextRC5;
end;
xa=uint8(RC5enctext);

function s=keygen(key,r)
% RC5/RC6 Key Generation
p=[99 81 225 183];
q=[185 121 55 158];
s(1,:)=p;
for i=2:2*r+4 s(i,:)=add(s(i-1,:),q);
end; i=1; a=zeros(1,4);
b=zeros(1,4); v=3*(2*r+4); for
h=1:v
  s(i,:)=add(s(i,:),a); s(i,:)=add(s(i,:),b);
  s(i,:)=shifting(s(i,:),3); a=s(i,:);
  key=add(key,a); key=add(key,b);
  key=shifting(key,LSB5(add(a,b)));
  b=key;
  if i==2*r+4 i=1;
  end end

function RC5enctextRC5=RC5OFB(plaintext1,m,n,CO,r,keyi)
%Encrypt Data using RC5 in OFB mode
RC5enctextRC5=zeros(m*n/8,8);
for i=1:m*n/8 plaintext=plaintext1(i,:);
  if i==1 cr=RC5enc(CO,r,keyi);
```

```
  else cr=RC5enc(cr,r,keyi);
  end; RC5enctext=bitxor(plaintext,cr);
  RC5enctextRC5(i,:)=RC5enctext;
end;

function plaintextRC5=RC5OFBDec(RC5enctext1,m,n,CO,r,keyi)
%Decrypt data using RC5 in OFB mode
plaintextRC5=zeros(m*n/8,8);
for i=1:m*n/8
  RC5enctextss=RC5enctext1(i,:);
  if i==1
    cr=RC5enc (CO,r,keyi);
    re_plaintext=bitxor(RC5enctextss,cr);
  else cr=RC5enc(cr,r,keyi);
    re_plaintext=bitxor(RC5enctextss,cr);
  end;
  plaintextRC5(i,:)=re_plaintext;
end;

function y=RC5sub(x,w)
%z=a-a1 mod 2^32
a=frombase256(x); a1=frombase256(w);
z=mod(a-a1,2^32); y=tobase256(z);

function RC6enctextRC6=CBC(plaintext1,m,n,CO,r,keyi)
%Encrypt Data using RC6 in CBC mode
RC6enctextRC6=zeros(m*n/16,16);
for i=1:m*n/16 plaintext=plaintext1(i,:); if i==1
    plaintexts=bitxor(plaintext,CO);
  else plaintexts=bitxor(plaintext,RC6enctext);
  end;
  RC6enctext = RC6enc(plaintexts,r,keyi);
  RC6enctextRC6(i,:)=RC6enctext;
end;

function plaintextRC6=CBCDec(RC6enctext1,m,n,CO,r,keyi)
%Decrypt data using RC6 in CBC mode
plaintextRC6=zeros(m*n/16,16);
for i=1:m*n/16
  RC6enctextss=RC6enctext1(i,:);
  re_plaintext = RC6dec(RC6enctextss,r,keyi);
  if i==1 plaintext=bitxor(re_plaintext,CO);
  else
    plaintext=bitxor(re_plaintext,RC6enctext1((i-1),:));
  end;
  plaintextRC6(i,:)=plaintext;
end;
```

```
function RC6enctextRC6=CFB(plaintext1,m,n,CO,r,keyi)
%Encrypt data using RC6 in CFB mode
RC6enctextRC6=zeros(m*n/16,16);
for i=1:m*n/16 plaintext=plaintext1(i,:); if i==1
    cr1=RC6enc(CO,r,keyi);
    RC6enctext=bitxor(plaintext,cr1);
  else cr1=RC6enc(RC6enctext,r,keyi);
    RC6enctext=bitxor(plaintext,cr1); end;
  RC6enctextRC6(i,:)=RC6enctext;
end;

function plaintextRC6=CFBDec(RC6enctext1,m,n,CO,r,keyi)
%Decrypt data using RC6 in CFB mode
plaintextRC6=zeros(m*n/16,16);
for i=1:m*n/16
  RC6enctextss=RC6enctext1(i,:);
  if i==1 cr1=RC6enc(CO,r,keyi);
    re_plaintext=bitxor(RC6enctextss,cr1);
  else
    cr1=RC6enc(RC6enctext1(i-1,:),r,keyi);
    re_plaintext=bitxor(RC6enctextss,cr1);
  end;
  plaintextRC6(i,:)=re_plaintext;
end;

function y=RC6dec(plaintext,round,s)
% RC6 Decryption
a=plaintext(1:4); b=plaintext(5:8);
c=plaintext(9:12); d=plaintext(13:16);
c=RC6sub(c,s(2*round+4,:));
a=RC6sub(a,s(2*round+3,:)); for i=round:-1:1
temp=d; d=c; c=b; b=a;
a=temp;
u=multi(d,(2*d+1)); u=shifting(u,5);
t=multi(b,(2*b+1)); t=shifting(t,5);
c=RC6sub(c,s(2*i+2,:)); c=shifting(c,-
LSB5(t)); c=bitxor(c,u);
a=RC6sub(a,s(2*i+1,:)); a=shifting(a,-
LSB5(u)); a=bitxor(a,t);
end d=RC6sub(d,s(2,:));
b=RC6sub(b,s(1,:)); y(1:4)=a;
y(5:8)=b; y(9:12)=c; y(13:16)=d;

function [xa1,Fs,nbits]=RC6DecCBCAudio(x1, Fs,
nbits,key,r,CO)
%Decrypt Audio using RC6 in CBC mode
%CBC keyi=RC6keygen(key,r); L=length(x1);
y1=double(x1); y1=reshape(y1,L/16,16);
```

```
plaintextRC6=zeros(L/16,16);
for i=1:L/16
RC6enctextss=y1(i,:);
re_plaintext = RC6dec(RC6enctextss,r,keyi);
if i==1
    plaintext=bitxor(re_plaintext,CO);
  else
    plaintext=bitxor(re_plaintext,y1((i-1),:));
  end;
  plaintextRC6(i,:)=plaintext; end;
xa1=uint8(plaintextRC6(:));

function xa1=RC6DecCBCImage(x1,key,r,CO)
%Decrypt Image using RC6 in CBC mode
keyi=RC6keygen(key,r);
%CBC [m,n]=size(x1);
y1=double(x1);
plaintextRC6=zeros(m*n/16,16);
RC6enctext1=reshape(y1',16,m*n/16)'; for i=1:m*n/16
  RC6enctextss=RC6enctext1(i,:);
  re_plaintext = RC6dec(RC6enctextss,r,keyi);
  if i==1
    plaintext=bitxor(re_plaintext,CO);
  else
    plaintext=bitxor(re_plaintext,RC6enctext1((i-1),:));
  end;
  plaintextRC6(i,:)=plaintext;
end;
ya1=reshape(plaintextRC6',n,m)';
xa1=uint8(ya1);

function xa1=RC6DecCBCImageC(x,key,r,CO)
%Decrypt Colored Image using RC6 in CBC mode
x1=x(:,:,1); x2=x(:,:,2); x3=x(:,:,3);
y1=double(x1); y2=double(x2);
y3=double(x3); [m,n]=size(x1);
plaintext(:,:,1)=reshape(y1',16,m*n/16)';
plaintext(:,:,2)=reshape(y2',16,m*n/16)';
plaintext(:,:,3)=reshape(y3',16,m*n/16)'; ciphertext=x;
keyi=RC6keygen(key,r);
%CBC
for i=1:3 plaintextRC6=plaintext(:,:,i);
  RC6enctext=RC6CBCDec(plaintextRC6,m,n,CO,r,keyi);
  RC6enctext=reshape(RC6enctext',n,m)';
  ciphertext(:,:,i)=RC6enctext;
end
xa1=uint8(ciphertext);
```

```
function [xa1, Fs, nbits]=RC6DecCFBAudio(x1, Fs,
nbits,key,r,CO)
%Decrypt Audio using RC6 in CFB mode
%CFB keyi=RC6keygen(key,r); L=length(x1);
y1=double(x1); y1=reshape(y1,L/16,16);
plaintextRC6=zeros(L/16,16);
for i=1:L/16
  RC6enctextss=y1(i,:);
  if i==1 cr1=RC6enc(CO,r,keyi);
    re_plaintext=bitxor(RC6enctextss,cr1);
  else
    cr1=RC6enc(y1(i-1,:),r,keyi);
    re_plaintext=bitxor(RC6enctextss,cr1);
  end;
  plaintextRC6(i,:)=re_plaintext;
end;
xa1=uint8(plaintextRC6(:));

function xa1=RC6DecCFBImage(x1,key,r,CO)
%Decrypt Image using RC6 in CFB mode
keyi=RC6keygen(key,r);
%CFB [m,n]=size(x1);
y1=double(x1);
plaintextRC6=zeros(m*n/16,16);
RC6enctext1=reshape(y1',16,m*n/16)'; for i=1:m*n/16
  RC6enctextss=RC6enctext1(i,:);
  if i==1 cr1=RC6enc(CO,r,keyi);
      re_plaintext=bitxor(RC6enctextss,cr1);
  else
    cr1=RC6enc(RC6enctext1(i-1,:),r,keyi);
      re_plaintext=bitxor(RC6enctextss,cr1);
  end;
  plaintextRC6(i,:)=re_plaintext;
end;
ya1=reshape(plaintextRC6',n,m)';
xa1=uint8(ya1);

function xa1=RC6DecCFBImageC(x,key,r,CO)
%Decrypt Colored Image using RC6 in CFB mode
x1=x(:,:,1); x2=x(:,:,2); x3=x(:,:,3);
y1=double(x1); y2=double(x2);
y3=double(x3); [m,n]=size(x1);
plaintext(:,:,1)=reshape(y1',16,m*n/16)';
plaintext(:,:,2)=reshape(y2',16,m*n/16)';
plaintext(:,:,3)=reshape(y3',16,m*n/16)';ciphertext=x;
keyi=RC6keygen(key,r);
%CFB
for i=1:3 plaintextRC6=plaintext(:,:,i);
  RC6enctext=RC6CFBDec(plaintextRC6,m,n,CO,r,keyi);
```

```
  RC6enctext=reshape(RC6enctext',n,m)';
  ciphertext(:,:,i)=RC6enctext;
end
xa1=uint8(ciphertext);

function [xa1, Fs, nbits]=RC6DecECBAudio(x1, Fs,
nbits,key,r)
%Decrypt Audio using RC6 in ECB mode
%ECB keyi=RC6keygen(key,r); L=length(x1);
y1=double(x1); y1=reshape(y1,L/16,16);
plaintextRC6=zeros(L/16,16);
for i=1:L/16
  RC6enctextss=y1(i,:);
  re_plaintext =RC6dec(RC6enctextss,r,keyi);
  plaintextRC6(i,:)=re_plaintext;
end;
xa1=uint8(plaintextRC6(:));

function xa1=RC6DecECBImage(x1,key,r)
%Decrypt Image using RC6 in ECB mode
keyi=RC6keygen(key,r);
%ECB [m,n]=size(x1);
y1=double(x1);
plaintextRC6=zeros(m*n/16,16);
RC6enctext1=reshape(y1',16,m*n/16)'; for i=1:m*n/16
  RC6enctextss=RC6enctext1(i,:);
  re_plaintext =RC6dec(RC6enctextss,r,keyi);
  plaintextRC6(i,:)=re_plaintext;
end;
ya1=reshape(plaintextRC6',n,m)';
xa1=uint8(ya1);

function xa1=RC6DecECBImageC(x,key,r)
%Decrypt Colored Image using RC6 in ECB mode
x1=x(:,:,1); x2=x(:,:,2); x3=x(:,:,3);
y1=double(x1); y2=double(x2);
y3=double(x3); [m,n]=size(x1);
plaintext(:,:,1)=reshape(y1',16,m*n/16)';
plaintext(:,:,2)=reshape(y2',16,m*n/16)';
plaintext(:,:,3)=reshape(y3',16,m*n/16)'; ciphertext=x;
keyi=RC6keygen(key,r);
%ECB
for i=1:3 plaintextRC6=plaintext(:,:,i);
  RC6enctext=RC6ECBDec(plaintextRC6,m,n,r,keyi);
  RC6enctext=reshape(RC6enctext',n,m)';
  ciphertext(:,:,i)=RC6enctext;
end xa1=uint8(ciphertext);
```

```
function [xa1, Fs, nbits]=RC6DecOFBAudio(x1, Fs,
nbits,key,r,CO)
%Decrypt Audio using RC6 in OFB mode
%OFB keyi=RC6keygen(key,r); L=length(x1);
y1=double(x1); y1=reshape(y1,L/16,16);
plaintextRC6=zeros(L/16,16);
for i=1:L/16
  RC6enctextss=y1(i,:);
  if i==1
    cr=RC6enc (CO,r,keyi);
    re_plaintext=bitxor(RC6enctextss,cr);
  else cr=RC6enc(cr,r,keyi);
    re_plaintext=bitxor(RC6enctextss,cr);
  end;
  plaintextRC6(i,:)=re_plaintext;
end;
xa1=uint8(plaintextRC6(:));

function xa1=RC6DecOFBImage(x1,key,r,CO)
%Decrypt Image using RC6 in OFB mode
keyi=RC6keygen(key,r);
%OFB [m,n]=size(x1);
y1=double(x1);
plaintextRC6=zeros(m*n/16,16);
RC6enctext1=reshape(y1',16,m*n/16)'; for i=1:m*n/16
  RC6enctextss=RC6enctext1(i,:);
  if i==1
      cr=RC6enc (CO,r,keyi);
    re_plaintext=bitxor(RC6enctextss,cr);
  else cr=RC6enc(cr,r,keyi);
    re_plaintext=bitxor(RC6enctextss,cr);
  end;
  plaintextRC6(i,:)=re_plaintext;
end;
ya1=reshape(plaintextRC6',n,m)';
xa1=uint8(ya1);

function xa1=RC6DecOFBImageC(x,key,r,CO)
%Decrypt Colored Image using RC6 in OFB mode
x1=x(:,:,1); x2=x(:,:,2); x3=x(:,:,3);
y1=double(x1); y2=double(x2);
y3=double(x3); [m,n]=size(x1);
plaintext(:,:,1)=reshape(y1',16,m*n/16)';
plaintext(:,:,2)=reshape(y2',16,m*n/16)';
plaintext(:,:,3)=reshape(y3',16,m*n/16)'; ciphertext=x;
keyi=RC6keygen(key,r);
%OFB
for i=1:3 plaintextRC6=plaintext(:,:,i);
  RC6enctext=RC6OFBDec(plaintextRC6,m,n,CO,r,keyi);
```

```
  RC6enctext=reshape(RC6enctext',n,m)';
  ciphertext(:,:,i)=RC6enctext;
end
xa1=uint8(ciphertext);

function RC6enctextRC6=RC6ECB(plaintext1,m,n,r,keyi)
%Encrypt data using RC6 in ECB mode
RC6enctextRC6=zeros(m*n/16,16);
for i=1:m*n/16 plaintext=plaintext1(i,:);
  RC6enctext = RC6enc(plaintext,r,keyi);
  RC6enctextRC6(i,:)=RC6enctext;
end;

function plaintextRC6=RC6ECBDec(RC6enctext1,m,n,r,keyi)
%Decrypt Data using RC6 in ECB mode
plaintextRC6=zeros(m*n/16,16);
for i=1:m*n/16
  RC6enctextss=RC6enctext1(i,:);
  re_plaintext =RC6dec(RC6enctextss,r,keyi);
  plaintextRC6(i,:)=re_plaintext;
end;

function y=RC6enc(plaintext,round,s)
%RC6 Encryption
a=plaintext(1:4); b=plaintext(5:8);
c=plaintext(9:12); d=plaintext(13:16);
b=add(b,s(1,:)); d=add(d,s(2,:));
  for i=1:round t=multi(b,(2*b+1)); t=shifting(t,5);
  u=multi(d,(2*d+1)); u=shifting(u,5);
  a=bitxor(a,t'); a=shifting(a,LSB5(u));
  a=add(a,s(2*i+1,:)); c=bitxor(c,u');
  c=shifting(c,LSB5(t)); c=add(c,s(2*i+2,:));
  temp=a;
  a=b; b=c; c=d; d=temp;
end a=add(a,s(2*round+3,:));
c=add(c,s(2*round+4,:)); y(1:4)=a;
y(5:8)=b; y(9:12)=c; y(13:16)=d;

function [xa, Fs, nbits]=RC6EncCBCAudio(x, Fs,
nbits,key,r,CO)
%Encrypt Audio using RC6 in CBC mode
L=length(x);
y=double(x); keyi=RC6keygen(key,r);
plaintext1=reshape(y,L/16,16);
RC6enctextRC6=zeros(L/16,16);
        %CBC
  for i=1:L/16
  plaintext=plaintext1(i,:);
  if i==1
  plaintexts=bitxor(plaintext,CO);
```

```
    else plaintexts=bitxor(plaintext,RC6enctext);
    end;
    RC6enctext = RC6enc(plaintexts,r,keyi);
    RC6enctextRC6(i,:)=RC6enctext;
    end;
  xa=uint8(RC6enctextRC6(:));

  function xa=RC6EncCBCImage(x,key,r,CO)
  %Encrypt Image using RC6 in CBC mode
  [m,n]=size(x);
  y=double(x);
  keyi=RC6keygen(key,r);
  plaintext1=reshape(y',16,m*n/16)';
  RC6enctextRC6=zeros(m*n/16,16);
    %CBC
  for i=1:m*n/16
    plaintext=plaintext1(i,:);
    if i==1
      plaintexts=bitxor(plaintext,CO);
    else
      plaintexts=bitxor(plaintext,RC6enctext);
    end;
    RC6enctext = RC6enc(plaintexts,r,keyi);
    RC6enctextRC6(i,:)=RC6enctext;
  end;
  ya=reshape(RC6enctextRC6',n,m)';
  xa=uint8(ya);

  function xa=RC6EncCBCImageC(x,key,r,CO)
  %Encrypt Colored Image using RC6 in CBC mode
  x1=x(:,:,1);
  x2=x(:,:,2);
  x3=x(:,:,3);
  [m,n]=size(x1);
  y1=double(x1);
  y2=double(x2);
  y3=double(x3);
  keyi=RC6keygen(key,r);
  plaintext(:,:,1)=reshape(y1',16,m*n/16)';
  plaintext(:,:,2)=reshape(y2',16,m*n/16)';
  plaintext(:,:,3)=reshape(y3',16,m*n/16)';
  RC6enctext=x;
    %CBC
  for i=1:3 plaintext1=plaintext(:,:,i);
    RC6enctextRC6=RC6CBC(plaintext1,m,n,CO,r,keyi);
    RC6enctextRC6=reshape(RC6enctextRC6',n,m)';
    RC6enctext(:,:,i)=RC6enctextRC6;
  end;
  xa=uint8(RC6enctext);
```

```
function [xa, Fs, nbits]=RC6EncCFBAudio(x, Fs,
nbits,key,r,CO)
%Encrypt Audio using RC6 in CFB mode
L=length(x);
y=double(x);
keyi=RC6keygen(key,r);
plaintext1=reshape(y,L/16,16);
RC6enctextRC6=zeros(L/16,16);
  % CFB
for i=1:L/16
  plaintext=plaintext1(i,:);
  if i==1
    cr1=RC6enc(CO,r,keyi);
    RC6enctext=bitxor(plaintext,cr1);
  else
    cr1=RC6enc(RC6enctext,r,keyi);
    RC6enctext=bitxor(plaintext,cr1);
  end;
  RC6enctextRC6(i,:)=RC6enctext;
end;
xa=uint8(RC6enctextRC6(:));

function xa=RC6EncCFBImage(x,key,r,CO)
%Encrypt Image using RC6 in CFB mode
[m,n]=size(x);
y=double(x);
keyi=RC6keygen(key,r);
plaintext1=reshape(y',16,m*n/16)';
RC6enctextRC6=zeros(m*n/16,16);
% CFB
for i=1:m*n/16
  plaintext=plaintext1(i,:);
  if i==1
    cr1=RC6enc(CO,r,keyi);
    RC6enctext=bitxor(plaintext,cr1);
  else
    cr1=RC6enc(RC6enctext,r,keyi);
    RC6enctext=bitxor(plaintext,cr1);
  end;
  RC6enctextRC6(i,:)=RC6enctext;
  end;
ya=reshape(RC6enctextRC6',n,m)';
xa=uint8(ya);

function xa=RC6EncCFBImageC(x,key,r,CO)
%Encrypt Colored Image using RC6 in CFB mode
x1=x(:,:,1);
x2=x(:,:,2);
x3=x(:,:,3);
[m,n]=size(x1);
```

```
y1=double(x1);
y2=double(x2);
y3=double(x3);
keyi=RC6keygen(key,r);
plaintext(:,:,1)=reshape(y1',16,m*n/16)';
plaintext(:,:,2)=reshape(y2',16,m*n/16)';
plaintext(:,:,3)=reshape(y3',16,m*n/16)';
RC6enctext=x;
% CFB
for i=1:3 plaintext1=plaintext(:,:,i);
  RC6enctextRC6=RC6CFB(plaintext1,m,n,CO,r,keyi);
  RC6enctextRC6=reshape(RC6enctextRC6',n,m)';
  RC6enctext(:,:,i)=RC6enctextRC6;
end;
xa=uint8(RC6enctext);

function [xa, Fs, nbits]=RC6EncECBAudio(x, Fs,
nbits,key,r)
%Encrypt Audio using RC6 in ECB mode
L=length(x);
y=double(x);
keyi=RC6keygen(key,r);
plaintext1=reshape(y,L/16,16);
RC6enctextRC6=zeros(L/16,16);
  % ECB
for i=1:L/16
  plaintext=plaintext1(i,:);
  RC6enctext = RC6enc(plaintext,r,keyi);
  RC6enctextRC6(i,:)=RC6enctext;
end;
xa=uint8(RC6enctextRC6(:));

function xa=RC6EncECBImage(x,key,r)
%Encrypt Image using RC6 in ECB mode
[m,n]=size(x);
y=double(x);
keyi=RC6keygen(key,r);
plaintext1=reshape(y',16,m*n/16)';
RC6enctextRC6=zeros(m*n/16,16);
% ECB
for i=1:m*n/16
  plaintext=plaintext1(i,:);
  RC6enctext = RC6enc(plaintext,r,keyi);
  RC6enctextRC6(i,:)=RC6enctext;
end;
ya=reshape(RC6enctextRC6',n,m)';
xa=uint8(ya);

function xa=RC6EncECBImageC(x,key,r)
%Encrypt Colored Image using RC6 in ECB mode
x1=x(:,:,1);
```

```
x2=x(:,:,2);
x3=x(:,:,3);
[m,n]=size(x1);
y1=double(x1);
y2=double(x2);
y3=double(x3);
keyi=RC6keygen(key,r);
plaintext(:,:,1)=reshape(y1',16,m*n/16)';
plaintext(:,:,2)=reshape(y2',16,m*n/16)';
plaintext(:,:,3)=reshape(y3',16,m*n/16)';
RC6enctext=x;
% ECB
for i=1:3 plaintext1=plaintext(:,:,i);
  RC6enctextRC6=RC6ECB(plaintext1,m,n,r,keyi);
  RC6enctextRC6=reshape(RC6enctextRC6',n,m)';
  RC6enctext(:,:,i)=RC6enctextRC6;
end;
xa=uint8(RC6enctext);

function [xa, Fs, nbits]=RC6EncOFBAudio(x, Fs,
nbits,key,r,CO)
%Encrypt Audio using RC6 in OFB mode
L=length(x);
y=double(x);
keyi=RC6keygen(key,r);
plaintext1=reshape(y,L/16,16);
RC6enctextRC6=zeros(L/16,16);
  % OFB
  for i=1:L/16
  plaintext=plaintext1(i,:);
   if i==1
    cr=RC6enc(CO,r,keyi);
   else
    cr=RC6enc(cr,r,keyi);
   end;
   RC6enctext=bitxor(plaintext,cr);
   RC6enctextRC6(i,:)=RC6enctext;
end;
xa=uint8(RC6enctextRC6(:));

function xa=RC6EncOFBImage(x,key,r,CO)
%Encrypt Image using RC6 in OFB mode
[m,n]=size(x);
y=double(x);
keyi=RC6keygen(key,r);
plaintext1=reshape(y',16,m*n/16)';
RC6enctextRC6=zeros(m*n/16,16);
% OFB
for i=1:m*n/16
  plaintext=plaintext1(i,:);
```

```
    if i==1
      cr=RC6enc(CO,r,keyi);
    else
      cr=RC6enc(cr,r,keyi);
    end;
    RC6enctext=bitxor(plaintext,cr);
    RC6enctextRC6(i,:)=RC6enctext;
end;
ya=reshape(RC6enctextRC6',n,m)';
xa=uint8(ya);

function xa=RC6EncOFBImageC(x,key,r,CO)
%Encrypt Colored Image using RC6 in OFB mode
x1=x(:,:,1);
x2=x(:,:,2);
x3=x(:,:,3);
[m,n]=size(x1);
y1=double(x1);
y2=double(x2);
y3=double(x3);
keyi=RC6keygen(key,r);
plaintext(:,:,1)=reshape(y1',16,m*n/16)';
plaintext(:,:,2)=reshape(y2',16,m*n/16)';
plaintext(:,:,3)=reshape(y3',16,m*n/16)';
RC6enctext=x;
% OFB
for i=1:3 plaintext1=plaintext(:,:,i);
  RC6enctextRC6=RC6OFB(plaintext1,m,n,CO,r,keyi);
  RC6enctextRC6=reshape(RC6enctextRC6',n,m)';
  RC6enctext(:,:,i)=RC6enctextRC6;
end;
xa=uint8(RC6enctext);

function s=keygen(key,r)
%RC6 Key Generation
p=[99 81 225 91];
q=[185 121 55 158];
s(1,:)=p;
for i=2:2*r+4
  s(i,:)=add(s(i-1,:),q);
end; i=1;
a=zeros(1,4);
b=zeros(1,4);
v=3*(2*r+4);
for h=1:v
s(i,:)=add(s(i,:),a);
  s(i,:)=add(s(i,:),b);
  s(i,:)=shifting(s(i,:),3);
  a=s(i,:);
  key=add(key,a);
```

```
  key=add(key,b);
  key=shifting(key,LSB5(add(a,b)));
  b=key;
  if i==2*r+4
    i=1;
  end
end

function RC6enctextRC6=OFB(plaintext1,m,n,CO,r,keyi)
%Encrypt data using RC6 in OFB mode
RC6enctextRC6=zeros(m*n/16,16);
for i=1:m*n/16
  plaintext=plaintext1(i,:);
  if i==1
    cr=RC6enc(CO,r,keyi);
  else
    cr=RC6enc(cr,r,keyi);
end;
RC6enctext=bitxor(plaintext,cr);
RC6enctextRC6(i,:)=RC6enctext;
end;

function plaintextRC6=RC6OFBDec(RC6enctext1,m,n,CO,r,
keyi)
%Decrypt data using RC6 in OFB mode
plaintextRC6=zeros(m*n/16,16);
for i=1:m*n/16
  RC6enctextss=RC6enctext1(i,:);
  if i==1
  cr=RC6enc (CO,r,keyi);
    re_plaintext=bitxor(RC6enctextss,cr);
  else
  cr=RC6enc(cr,r,keyi);
  re_plaintext=bitxor(RC6enctextss,cr);
  end;
  plaintextRC6(i,:)=re_plaintext;
end;

function y=RC6sub(x,w)
%z=a-a1mod 2^32
a=frombase256(x);
a1=frombase256(w);
z=mod(a-a1,2^32);
y=tobase256(z);

function out = shifting (w,n)
% shifting w by n
y=dec2bin(w,8);
y=rot90 (y); y=y(:);
y=circshift(y,n);
```

```
y=reshape(y,8,4)';
out=bin2dec(y(:,end:-1:1));

function y=shiftleft(key,round)
% RC6 Shifting
if (round==1||round==2||round==9||round==16)
y=circshift(key,1); else
y=circshift(key,2); end

function y=tobase256(x)
%convert Hex to 256 bits base
for i=4:-1:1
  y(i)=fix(x/256^(i-1)); x=x-(y(i)*256^(i-1));
end

function c=add(a,b)
%c=a+b mode 2^8
c=zeros(1,4);
for i=1:4
  c(i)=mod(a(i)+b(i),2^8); if (i+1)~=5
    a(i+1)=a(i+1)+fix((a(i)+b(i))/256);
    end;
    end;

function y=binvec2decA(x)
%convert binary to decimall
y=0;
for i=1:length(x)
  y=y+x(i)*2^(i-1);
end;

function cipherdes1=cipher(plaindes1,r,keyi)
%%%%%%%%%%%%%%%%%%%%%%%%%%%%%%%%%%%%%%%%%%%%%%%%%%%%%%%%%%%%%%%%%%%
%this file for the encryption of the plaintext to a
ciphertext using des
% clc
%
clear
all
bin=zeros(8);
for i=1:8
  bin(i,:)=dec2binvecA(plaindes1(i),8);
end;
plaindes=rot90(bin);
plaindes=plaindes(:);
ip=InitialPermutation(plaindes);
left=ip(1:32);
right=ip(33:64);
for round=1:r
  expantion=exp1(right);
  xor_one=bitxor(expantion',keyi(round,:));
```

```
  substitution=DESsub(xor_one);
  permutation=p(substitution); xor_
  two=bitxor(left,permutation);
if round ~=r left=right;
  right=xor_two;
else
  left=xor_two;
end;
end;
y(1:32)=left; y(33:64)=right ;
cipherdesBIN1=finalpermutation(y);
cipherdesBIN1=reshape(cipherdesBIN1,8,8)';
cipherdes1=zeros(1,8);
for i=1:8
  cipherdes1(i)=binvec2decA(cipherdesBIN1(i,end:-1:1));
end;

function y=Convert256toHex(x)
%convert 256 base numbers to Hex.
a=length(x);
j=1;
y=zeros(1,2*a);
for i=1:a
y(j)=fix(x(i)/16); y(j+1)=x(i)-y(j)*16; j=j+2;
end

function y=dec2binvecA(x,size)
%Convert decimal to binary
y=zeros(1,size);
z=zeros(1,size+1);
z(1)=x;
for i=1:size
  y(i)=mod(z(i),2^i)/2^(i-1);
  z(i+1)=z(i)-y(i)*2^(i-1);
end;
% y=y(end:-1:1);

function plaindes1=decry(cipherdes1,r,keyi)
%DES decryption
bin=zeros(8);
for i=1:8
    bin(i,:)=dec2binvecA(cipherdes1(i),8);
end;
cipherdes=rot90(bin);
cipherdes=cipherdes(:);
ip=InitialPermutation(cipherdes);
 left=ip(1:32);
 right=ip(33:64);
for round=r:-1:1 expantion=exp1(right);
```

```
  xor_one=bitxor(expantion',keyi(round,:));
  substitution=DESsub(xor_one);
  permutation=p(substitution);
  xor_two=bitxor(left,permutation);
if round ~=1
  left=right;
  right=xor_two;
else
  left=xor_two;
end;
end;
 y(1:32)=left;
 y(33:64)=right ;
plaindesBIN1=finalpermutation(y);
plaindesBIN1=reshape(plaindesBIN1,8,8)';
plaindes1=zeros(1,8);
for i=1:8
  plaindes1(i)=binvec2decA(plaindesBIN1(i,end:-1:1));
end;

function DESenctextDES=DESCBC(plaintext1,m,n,CO,r,keyi)
%Encrypt data using DES in CBC mode
DESenctextDES=zeros(m*n/8,8);
for i=1:m*n/8
  plaintext=plaintext1(i,:);
  if i==1
    plaintexts=bitxor(plaintext,CO);
  else
    plaintexts=bitxor(plaintext,DESenctext);
  end;
  DESenctext = cipher(plaintexts,r,keyi);
  DESenctextDES(i,:)=DESenctext;
end;

function plaintextDES=DESCBCDec(DESenctext1,m,n,
CO,r,keyi)
%Decrypt data using DES in CBC mode
plaintextDES=zeros(m*n/8,8);
for i=1:m*n/8
  DESenctextss=DESenctext1(i,:);
  re_plaintext = decry(DESenctextss,r,keyi);
  if i==1
    plaintext=bitxor(re_plaintext,CO);
else
plaintext=bitxor(re_plaintext,DESenctext1((i-1),:));
end;
plaintextDES(i,:)=plaintext;
end;
```

```
function ciphertextDES=DESCFB(plaintext1,m,n,CO,r,keyi)
%Encrypt data using DES in CFB mode
ciphertextDES=zeros(m*n/8,8);
for i=1:m*n/8
  plaintext=plaintext1(i,:);
  if i==1
    cr1=cipher(CO,r,keyi);
    ciphertext=bitxor(plaintext,cr1);
  else
    cr1=cipher(ciphertext,r,keyi);
    ciphertext=bitxor(plaintext,cr1);
  end;
  ciphertextDES(i,:)=ciphertext;
end;

function plaintextDES=DESCFBDec(ciphertext1,m,n,
CO,r,keyi)
%decrypt data using DES in CFB mode
plaintextDES=zeros(m*n/8,8);
for i=1:m*n/8
  ciphertextss=ciphertext1(i,:);
  if i==1
    cr1=cipher(CO,r,keyi);
    re_plaintext=bitxor(ciphertextss,cr1);
  else
    cr1=cipher(ciphertext1(i-1,:),r,keyi);
    re_plaintext=bitxor(ciphertextss,cr1);
  end;
  plaintextDES(i,:)=re_plaintext;
end;

function [xa1,Fs,nbits]=DesDecCBCAudio(x1,Fs,nbits,
key,r,CO)
%Decryption Audio using DES in CBC mode
keyi=DESkeygen(key,r);
%CBC L=length(x1);
y1=double(x1);
plaintextDES=zeros(L/8,8);
ciphertext1=reshape(y1,L/8,8);
for i=1:L/8
  ciphertextss=ciphertext1(i,:);
  re_plaintext = decry(ciphertextss,r,keyi);
  if i==1
    plaintext=bitxor(re_plaintext,CO);
  else
    plaintext=bitxor(re_plaintext,ciphertext1((i-1),:));
  end;
  plaintextDES(i,:)=plaintext;
end;
xa1=uint8(plaintextDES(:));
```

```
function xa1=DesDecCBCImage(x1,key,r,CO)
%Decrypt Image using DES in CBC mode
keyi=DESkeygen(key,r);
%CBC
[m,n]=size(x1);
y1=double(x1);
plaintextDES=zeros(m*n/8,8);
ciphertext1=reshape(y1',8,m*n/8)';
for i=1:m*n/8
  ciphertextss=ciphertext1(i,:);
  re_plaintext = decry(ciphertextss,r,keyi);
  if i==1
    plaintext=bitxor(re_plaintext,CO);
  else
    plaintext=bitxor(re_plaintext,ciphertext1((i-1),:));
  end;
  plaintextDES(i,:)=plaintext;
end;
ya1=reshape(plaintextDES',n,m)';
xa1=uint8(ya1);

function xa1=DesDecCBCImageC(x,key,r,CO)
%Decrypt Colored Image using DES in CBC mode
keyi=DESkeygen(key,r);
%CBC
x1=x(:,:,1);
x2=x(:,:,2); x3=x(:,:,3);
y1=double(x1);
y2=double(x2);
y3=double(x3);
[m,n]=size(x1);
plaintext(:,:,1)=reshape(y1',8,m*n/8)';
plaintext(:,:,2)=reshape(y2',8,m*n/8)';
plaintext(:,:,3)=reshape(y3',8,m*n/8)';
ciphertext=x;
for i=1:3
plaintextDES=plaintext(:,:,i);
DESenctext=DESCBCDec(plaintextDES,m,n,CO,r,keyi);
DESenctext=reshape(DESenctext',n,m)';
ciphertext(:,:,i)=DESenctext;
end;
a1=uint8(ciphertext);

function [xa1,Fs,nbits]=DesDecCFBAudio(x1,Fs,nbits,
key,r,CO)
%Decrypt Audio using DES in CFB mode
keyi=DESkeygen(key,r);
%CFB L=length(x1); y1=double(x1);
plaintextDES=zeros(L/8,8);
```

```
ciphertext1=reshape(y1,L/8,8);
for i=1:L/8
  ciphertextss=ciphertext1(i,:);
  if i==1
    cr1=cipher(CO,r,keyi);
    re_plaintext=bitxor(ciphertextss,cr1);
  else
    cr1=cipher(ciphertext1(i-1,:),r,keyi);
    re_plaintext=bitxor(ciphertextss,cr1);
  end;
  plaintextDES(i,:)=re_plaintext;
end;
xa1=uint8(plaintextDES(:));

function xa1=DesDecCFBImage(x1,key,r,CO)
%Decrypt Image using DES in CFB mode
keyi=DESkeygen(key,r);
%CFB
[m,n]=size(x1);
y1=double(x1);
plaintextDES=zeros(m*n/8,8);
ciphertext1=reshape(y1',8,m*n/8)';
for i=1:m*n/8
  ciphertextss=ciphertext1(i,:);
  if i==1
    cr1=cipher(CO,r,keyi);
    re_plaintext=bitxor(ciphertextss,cr1);
  else
    cr1=cipher(ciphertext1(i-1,:),r,keyi);
    re_plaintext=bitxor(ciphertextss,cr1);
  end;
  plaintextDES(i,:)=re_plaintext;
end;
ya1=reshape(plaintextDES',n,m)';
xa1=uint8(ya1);

function xa1=DesDecCFBImageC(x,key,r,CO)
%Decrypt Colored Image using DES in CFB mode
keyi=DESkeygen(key,r);
%CFB
x1=x(:,:,1);
x2=x(:,:,2);
x3=x(:,:,3);
y1=double(x1);
y2=double(x2);
y3=double(x3);
[m,n]=size(x1);
plaintext(:,:,1)=reshape(y1',8,m*n/8)';
plaintext(:,:,2)=reshape(y2',8,m*n/8)';
```

```
plaintext(:,:,3)=reshape(y3',8,m*n/8)';
ciphertext=x;
for i=1:3 plaintextDES=plaintext(:,:,i);
  DESenctext=DESCFBDec(plaintextDES,m,n,CO,r,keyi);
  DESenctext=reshape(DESenctext',n,m)';
  ciphertext(:,:,i)=DESenctext;
end
xa1=uint8(ciphertext);

function [xa1,Fs,nbits]=DesDecECBAudio(x1,Fs,nbits,key,r)
%Decrypt Audio using DES in ECB mode
keyi=DESkeygen(key,r);
%ECB L=length(x1);
y1=double(x1);
plaintextDES=zeros(L/8,8);
ciphertext1=reshape(y1,L/8,8);
for i=1:L/8 ciphertextss=ciphertext1(i,:);
  re_plaintext =decry(ciphertextss,r,keyi);
  plaintextDES(i,:)=re_plaintext;
end;
xa1=uint8(plaintextDES(:));

function xa1=DesDecECBImage(x1,key,r)
%Decrypt Image using DES in ECB mode
keyi=DESkeygen(key,r);
%ECB
[m,n]=size(x1);
y1=double(x1);
plaintextDES=zeros(m*n/8,8);
ciphertext1=reshape(y1',8,m*n/8)';
for i=1:m*n/8
  ciphertextss=ciphertext1(i,:);
  re_plaintext =decry(ciphertextss,r,keyi);
  plaintextDES(i,:)=re_plaintext;
end;
ya1=reshape(plaintextDES',n,m)';
xa1=uint8(ya1);

function xa1=DesDecECBImageC(x,key,r)
%Decrypt Colored Image using DES in ECB mode
keyi=DESkeygen(key,r);
%ECB
x1=x(:,:,1);
x2=x(:,:,2);
x3=x(:,:,3);
y1=double(x1);
y2=double(x2);
y3=double(x3);
[m,n]=size(x1);
plaintext(:,:,1)=reshape(y1',8,m*n/8)';
```

```
plaintext(:,:,2)=reshape(y2',8,m*n/8)';
plaintext(:,:,3)=reshape(y3',8,m*n/8)';
ciphertext=x;
for i=1:3 plaintextDES=plaintext(:,:,i);
  DESenctext=DESECBDec(plaintextDES,m,n,r,keyi);
  DESenctext=reshape(DESenctext',n,m)';
  ciphertext(:,:,i)=DESenctext;
end
xa1=uint8(ciphertext);

function [xa1,Fs,nbits]=DesDecOFBAudio(x1,Fs,nbits,
key,r,CO)
%Decrypt Audio using DES in OFB mode
keyi=DESkeygen(key,r);
%OFB L=length(x1);
y1=double(x1);
plaintextDES=zeros(L/8,8);
ciphertext1=reshape(y1,L/8,8);
for i=1:L/8
  ciphertextss=ciphertext1(i,:);
  if i==1
    cr=cipher(CO,r,keyi);
    re_plaintext=bitxor(ciphertextss,cr);
  else cr=cipher(cr,r,keyi);
    re_plaintext=bitxor(ciphertextss,cr);
  end;
  plaintextDES(i,:)=re_plaintext;
end;
xa1=uint8(plaintextDES(:));

function xa1=DesDecOFBImage(x1,key,r,CO)
%Decrypt Image using DES in OFB mode
keyi=DESkeygen(key,r);
%OFB
[m,n]=size(x1);
y1=double(x1);
plaintextDES=zeros(m*n/8,8);
ciphertext1=reshape(y1',8,m*n/8)';
for i=1:m*n/8
  ciphertextss=ciphertext1(i,:);
  if i==1 cr=cipher(CO,r,keyi);
    re_plaintext=bitxor(ciphertextss,cr);
  else cr=cipher(cr,r,keyi);
    re_plaintext=bitxor(ciphertextss,cr);
  end;
  plaintextDES(i,:)=re_plaintext;
end;
ya1=reshape(plaintextDES',n,m)';
xa1=uint8(ya1);
```

```
function xa1=DesDecOFBImageC(x,key,r,CO)
%Decrypt Colored Image using DES in OFB mode
keyi=DESkeygen(key,r);
%OFB
x1=x(:,:,1);
x2=x(:,:,2);
x3=x(:,:,3);
y1=double(x1);
y2=double(x2);
y3=double(x3);
[m,n]=size(x1);
plaintext(:,:,1)=reshape(y1',8,m*n/8)';
plaintext(:,:,2)=reshape(y2',8,m*n/8)';
plaintext(:,:,3)=reshape(y3',8,m*n/8)';
ciphertext=x;
for i=1:3 plaintextDES=plaintext(:,:,i);
  DESenctext=DESOFBDec(plaintextDES,m,n,CO,r,keyi);
  DESenctext=reshape(DESenctext',n,m)';
  ciphertext(:,:,i)=DESenctext;
end
xa1=uint8(ciphertext);

function DESenctextDES=DESECB(plaintext1,m,n,r,keyi)
%Encrypt Data using DES in ECB mode
DESenctextDES=zeros(m*n/8,8);
for i=1:m*n/8
  plaintext=plaintext1(i,:);
  DESenctext = cipher(plaintext,r,keyi);
  DESenctextDES(i,:)=DESenctext;
end;

function plaintextDES=DESECBDec(DESenctext1,m,n,r,keyi)
%Decrypt data using DES in ECB mode
plaintextDES=zeros(m*n/8,8);
for i=1:m*n/8
  DESenctextss=DESenctext1(i,:);
  re_plaintext =decry(DESenctextss,r,keyi);
  plaintextDES(i,:)=re_plaintext;
end;

function [xa,Fs,nbits]=DesEncCBCAudio(x,Fs,nbits,
key,r,CO)
%Encrypt Audio using DES in CBC mode
L=length(x);
y=double(x);
keyi=DESkeygen(key,r);
plaintext1=reshape(y,L/8,8);
ciphertextDES=zeros(L/8,8);
  %CBC
```

```
for i=1:L/8
  plaintext=plaintext1(i,:);
  if i==1
    plaintexts=bitxor(plaintext,CO);
  else
    plaintexts=bitxor(plaintext,ciphertext);
  end;
  ciphertext = cipher(plaintexts,r,keyi);
  ciphertextDES(i,:)=ciphertext;
end;
xa=uint8(ciphertextDES(:));

function xa=DesEncCBCImage(x,key,r,CO)
%Encrypt Image using DES in CBC mode
[m,n]=size(x);
y=double(x);
keyi=DESkeygen(key,r);
plaintext1=reshape(y',8,m*n/8)';
ciphertextDES=zeros(m*n/8,8);
  %CBC
for i=1:m*n/8
  plaintext=plaintext1(i,:);
  if i==1
    plaintexts=bitxor(plaintext,CO);
  else
    plaintexts=bitxor(plaintext,ciphertext);
  end;
  ciphertext = cipher(plaintexts,r,keyi);
  ciphertextDES(i,:)=ciphertext;
end;
ya=reshape(ciphertextDES',n,m)';
xa=uint8(ya);

function xa=DesEncCBCImageC(x,key,r,CO)
%Encrypt Colored Image using DES in CBC mode
x1=x(:,:,1);
x2=x(:,:,2);
x3=x(:,:,3);
[m,n]=size(x1);
y1=double(x1);
y2=double(x2);
y3=double(x3);
keyi=DESkeygen(key,r);
plaintext(:,:,1)=reshape(y1',8,m*n/8)';
plaintext(:,:,2)=reshape(y2',8,m*n/8)';
plaintext(:,:,3)=reshape(y3',8,m*n/8)';
DESenctext=x;
  %CBC
for i=1:3 plaintext1=plaintext(:,:,i);
```

```
  DESenctextDES=DESCBC(plaintext1,m,n,CO,r,keyi);
  DESenctextDES=reshape(DESenctextDES',n,m)';
  DESenctext(:,:,i)=DESenctextDES;
end;
xa=uint8(DESenctext);

function [xa,Fs,nbits]=DesEncCFBAudio(x,Fs,nbits,
key,r,CO)
%Encrypt Audio using DES in CFB mode
L=length(x);
y=double(x);
keyi=DESkeygen(key,r);
plaintext1=reshape(y,L/8,8);
ciphertextDES=zeros(L/8,8);
  % CFB
for i=1:L/8
  plaintext=plaintext1(i,:);
  if i==1
    cr1=cipher(CO,r,keyi);
    ciphertext=bitxor(plaintext,cr1);
  else cr1=cipher(ciphertext,r,keyi);
    ciphertext=bitxor(plaintext,cr1);
  end;
  ciphertextDES(i,:)=ciphertext;
end;
xa=uint8(ciphertextDES(:));

function xa=DesEncCFBimage(x,key,r,CO)
%Encrypt Image using DES in CFB mode
[m,n]=size(x);
y=double(x);
keyi=DESkeygen(key,r);
plaintext1=reshape(y',8,m*n/8)';
ciphertextDES=zeros(m*n/8,8);
% CFB
for i=1:m*n/8
  plaintext=plaintext1(i,:);
  if i==1
    cr1=cipher(CO,r,keyi);
    ciphertext=bitxor(plaintext,cr1);
  else cr1=cipher(ciphertext,r,keyi);
    ciphertext=bitxor(plaintext,cr1);
  end;
ciphertextDES(i,:)=ciphertext;
end;
ya=reshape(ciphertextDES',n,m)';
xa=uint8(ya);
```

```
function xa=DesEncCFBImageC(x,key,r,CO)
%Encrypt Colored Image using DES in CFB mode
x1=x(:,:,1); x2=x(:,:,2); x3=x(:,:,3);
[m,n]=size(x1); y1=double(x1);
y2=double(x2); y3=double(x3);
keyi=DESkeygen(key,r);
plaintext(:,:,1)=reshape(y1',8,m*n/8)';
plaintext(:,:,2)=reshape(y2',8,m*n/8)';
plaintext(:,:,3)=reshape(y3',8,m*n/8)'; DESenctext=x;
  % CFB
for i=1:3 plaintext1=plaintext(:,:,i);
  DESenctextDES=DESCFB(plaintext1,m,n,CO,r,keyi);
  DESenctextDES=reshape(DESenctextDES',n,m)';
  DESenctext(:,:,i)=DESenctextDES;
end;
xa=uint8(DESenctext);

function [xa,Fs,nbits]=DesEncECBAudio(x,Fs,nbits,key,r)
%Encrypt Audio using DES in ECB mode
L=length(x);
y=double(x); keyi=DESkeygen(key,r);
plaintext1=reshape(y,L/8,8);
ciphertextDES=zeros(L/8,8);
  % ECB
for i=1:L/8 plaintext=plaintext1(i,:);
  ciphertext = cipher(plaintext,r,keyi);
  ciphertextDES(i,:)=ciphertext;
end;
xa=uint8(ciphertextDES(:));

function xa=DesEncECBImage(x,key,r)
%Encrypt Image using DES in ECB mode
[m,n]=size(x);
y=double(x); keyi=DESkeygen(key,r);
plaintext1=reshape(y',8,m*n/8)';
ciphertextDES=zeros(m*n/8,8);
% ECB
for i=1:m*n/8 plaintext=plaintext1(i,:);
  ciphertext = cipher(plaintext,r,keyi);
  ciphertextDES(i,:)=ciphertext;
end; ya=reshape(ciphertextDES',n,m)'; xa=uint8(ya);

function xa=DesEncECBImageC(x,key,r)
%Encrypt Colored Image using DES in ECB mode
x1=x(:,:,1); x2=x(:,:,2); x3=x(:,:,3);
[m,n]=size(x1); y1=double(x1);
y2=double(x2);
y3=double(x3); keyi=DESkeygen(key,r);
plaintext(:,:,1)=reshape(y1',8,m*n/8)';
```

```
plaintext(:,:,2)=reshape(y2',8,m*n/8)';
plaintext(:,:,3)=reshape(y3',8,m*n/8)'; DESenctext=x;
for i=1:3 plaintext1=plaintext(:,:,i);
  DESenctextDES=DESECB(plaintext1,m,n,r,keyi);
  DESenctextDES=reshape(DESenctextDES',n,m)';
  DESenctext(:,:,i)=DESenctextDES;
end;
xa=uint8(DESenctext);

function [xa,Fs,nbits]=DesEncOFBAudio(x,Fs,nbits,
key,r,CO)
%Encrypt Image using DES in OFB mode
L=length(x);
y=double(x); keyi=DESkeygen(key,r);
plaintext1=reshape(y,L/8,8);
ciphertextDES=zeros(L/8,8);
  % OFB
for i=1:L/8 plaintext=plaintext1(i,:); if i==1
    cr=cipher(CO,r,keyi);
  else cr=cipher(cr,r,keyi);
  end; ciphertext=bitxor(plaintext,cr);
  ciphertextDES(i,:)=ciphertext;
end;
xa=uint8(ciphertextDES(:));

function xa=DesEncOFBImage(x,key,r,CO)
%Encrypt Image using DES in OFB mode
[m,n]=size(x);
y=double(x); keyi=DESkeygen(key,r);
plaintext1=reshape(y',8,m*n/8)';
ciphertextDES=zeros(m*n/8,8);
% OFB
for i=1:m*n/8 plaintext=plaintext1(i,:); if i==1
    cr=cipher(CO,r,keyi);
  else cr=cipher(cr,r,keyi);
  end; ciphertext=bitxor(plaintext,cr);
  ciphertextDES(i,:)=ciphertext;
end; ya=reshape(ciphertextDES',n,m)'; xa=uint8(ya);

function xa=DesEncOFBImageC(x,key,r,CO)
%Encrypt Colored Image using DES in OFB mode
x1=x(:,:,1);
x2=x(:,:,2);
x3=x(:,:,3);
[m,n]=size(x1);
y1=double(x1);
y2=double(x2);
y3=double(x3);
keyi=DESkeygen(key,r);
plaintext(:,:,1)=reshape(y1',8,m*n/8)';
```

```
plaintext(:,:,2)=reshape(y2',8,m*n/8)';
plaintext(:,:,3)=reshape(y3',8,m*n/8)';
DESenctext=x;
% OFB
for i=1:3 plaintext1=plaintext(:,:,i);
  DESenctextDES=DESOFB(plaintext1,m,n,CO,r,keyi);
  DESenctextDES=reshape(DESenctextDES',n,m)';
  DESenctext(:,:,i)=DESenctextDES;
end;
xa=uint8(DESenctext);

function key1=DESkeygen(Mainkey,r)
%DES Key Generation
%Inserting master key
%clc
%clear all
%Mainkey=['A';'A';'B';'B';'0';'9';'1';'8';'2';'7';'3';
'6';'C';'C';'D';'D'];
key2=Convert256toHex(Mainkey);
key=zeros(r,4); for i=1:r
key(i,:)=dec2binvecA(key2(i),4); end
key=rot90(key);
key=key(:);
a=[57 49 41 33 25 17 9 1 58 50 42 34 26 18 10 2 59 51 43
35 27 19 11 3 60 52 44 36 63 55 47 39 31 23 15 7 62 54 46
38 30 22 14 6 61 53 45 37 29 21 13 5 28 20 12 4];
b=[14 17 11 24 1 5 3 28 15 6 21 10 23 19 12 4 26 8 16
7 27 20 13 2 41 52 31 37 47 55 30 40 51 45 33 48 44 49
39 56 34 53 46 42 50 36 29 32];
%%%%%%%%%%%%%%%%%%%%%%%%%%%%%%%%%%%%%%
output1=key(a);
key1=zeros(r,48);
left=output1(28:-1:1);
right=output1(56:-1:29);
for i=1:r
left=shiftleft(left,i);
right=shiftleft(right,i);
output(28:-1:1)=left;
output(56:-1:29)=right;
key1(i,:)=output(b);
end

function ciphertextDES=DESOFB(plaintext1,m,n,CO,r,keyi)
%Encrypt data using DES in OFB mode
ciphertextDES=zeros(m*n/8,8);
for i=1:m*n/8
  plaintext=plaintext1(i,:);
  if i==1
    cr=cipher(CO,r,keyi);
```

```
    else
      cr=cipher(cr,r,keyi);
    end;
    ciphertext=bitxor(plaintext,cr);
    ciphertextDES(i,:)=ciphertext;
end;

function plaintextDES=DESOFBDec(ciphertext1,m,n,
CO,r,keyi)
%Decrypt data using DES in OFB mode
plaintextDES=zeros(m*n/8,8);
for i=1:m*n/8
    ciphertextss=ciphertext1(i,:);
    if i==1
      cr=cipher (CO,r,keyi);
      re_plaintext=bitxor(ciphertextss,cr);
    else cr=cipher(cr,r,keyi);
      re_plaintext=bitxor(ciphertextss,cr);
    end;
    plaintextDES(i,:)=re_plaintext;
end;

function y=sub(in)
%DES substitution
in1=reshape(in',6,8)';
x=zeros(1,8);
for i=1:8
  x(i)=binvec2decA(in1(i,end:-1:1));
end;
s= [14 0 4 15 13 7 1 4 2 14 15 2 11 13 8 1 3 10 10 6 6
12 12 11 5 9 9 5 0 3 7 8 4 15 1 12 14 8 8 2 13 4 6 9 2
1 11 7 15 5 12 11 9 3 7 14 3 10 10 0 5 6 0 13;
15 3 1 13 8 4 14 7 6 15 11 2 3 8 4 14 9 12 7 0 2 1 13
10 12 6 0 9 5 11 10 5 0 13 14 8 7 10 11 1 10 3 4 15 13
4 1 2 5 11 8 6 12 7 6 12 9 0 3 5 2 14 15 9;
10 13 0 7 9 0 14 9 6 3 3 4 15 6 5 10 1 2 13 8 12 5 7 14
11 12 4 11 2 15 8 1 13 1 6 10 4 13 9 0 8 6 15 9 3 8 0
7 11 4 1 5 2 14 12 3 5 11 10 5 14 2 7 12;
7 13 13 8 14 11 3 5 0 6 6 15 9 0 10 3 1 4 2 7 8 2 5 12
11 1 12 10 4 14 15 9 10 3 6 15 9 0 0 6 12 10 11 1 7 13
13 8 15 9 1 4 3 5 14 11 5 12 2 7 8 2 4 14;
2 14 12 11 4 2 1 12 7 4 10 7 11 13 6 1 8 5 5 0 3 15 15
10 13 3 0 9 14 8 9 6 4 11 2 8 1 12 11 7 10 1 13 14 7 2
8 13 15 6 9 15 12 0 5 9 6 10 3 4 0 5 14 3;
12 10 1 15 10 4 15 2 9 7 2 12 6 9 8 5 0 6 13 1 3 13 4
14 14 0 7 11 5 3 11 8 9 4 14 3 15 2 5 12 2 9 8 5 12 15
3 10 7 11 0 14 4 1 10 7 1 6 13 0 11 8 6 13;
4 13 11 0 2 11 14 7 15 4 0 9 8 1 13 10 3 14 12 3 9 5 7
12 5 2 10 15 6 8 1 6 1 6 4 11 11 13 13 8 12 1 3 4 7 10
14 7 10 9 15 5 6 0 8 15 0 14 5 2 9 3 2 12;
```

```
13 1 2 15 8 13 4 8 6 10 15 3 11 7 1 4 10 12 9 5 3 6 14
11 5 0 0 14 12 9 7 2 7 2 11 1 4 14 1 7 9 4 12 10 14 8 2
13 0 15 6 12 10 9 13 0 15 3 3 5 5 6 8 11];
out1=zeros(1,8);
for i=1:8
out1(i) = s(i,x(i)+1);
end
out=zeros(8,4);
for i=1:8
  out(i,:)=dec2binvecA(out1(i),4);
end
out=rot90(out);
y=out(:);

function expantion=exp1(right)
%DES expansion
a=[32,1,2,3,4,5,4,5,6,7,8,9,8,9,10,11,12,13,12,13,14,1
5,16,17,16,17,18,19,20,21,20,21,22,23,24,25,24,25,26,2
7,28,29,28,29,30,31,32,1];
expantion=right(a);

function y=finalpermutation(x)
%DES final permutation
a=[40 8 48 16 56 24 64 32 39 7 47 15 55 23 63 31 38 6
46 14 54 22 62 30 37 5 45 13 53 21 61 29 36 4 44 12 52
20 60 28 35 3 43 11 51 19 59 27 34 2 42 10 50 18 58 26
33 1 41 9 49 17 57 25];
y=x(a);

function y=frombase256(x)
%Convert 256 base number to decimal
y=0;
for i=1:4
  y=y+(x(i)*256^(i-1));
end

function ip=InitialPermutation(plaindes)
%DES Initial Permutation
a=[58 50 42 34 26 18 10 2 60 52 44 36 28 20 12 4 62 54
46 38 30 22 14 6 64 56 48 40 32 24 16 8 57 49 41 33 25
17 9 1 59 51 43 35 27 19 11 3 61 53 45 37 29 21 13 5 63
55 47 39 31 23 15 7];
ip=plaindes(a);

function y=LSB5(x)
%return the 5 least significant bits of x
y=dec2bin(x(1),8);
u=y(8:-1:3);
u=u(end:-1:1);
y=bin2dec(u);
```

```
function y=CHaoticCipher(im)
% This function encrypts square image using baker map
im=double(im);
n = [10,5,12,5,10,8,14,10,5,12,5,10,8,14,10,5,12,5,10,
8,14,10,5,12,5,10,8,14,10,5,12,5,10,8,14,10,5,12,5,10
,8,14,10,5,12,5,10,8,14,10,5,12,5,10,8,14,];
[pr,pc] = chaomat(n);
pim = chaoperm(im,pr,pc,3,'forward');
y=uint8(pim);
imshow(y);

function [pr,pc]=chaomat(n)
%
I=sum(n);
k=size(n,2);
for i=1:k
  N(i+1)=1;
  for j=1:i
    N(i+1)=N(i+1)+n(j);
  end
end
N(1)=1;
for cb=1:k
  for rb=1:n(cb)
    rbstartcol(rb)=mod((rb-1)*I,n(cb));
    rbendcol(rb)=mod((rb*I-1),n(cb));
    rbstartrow(rb)=fix(((rb-1)*I)/n(cb));
    rbendrow(rb)=fix((rb*I-1)/n(cb));
    mincol(rb)=min([rbendcol(rb)+1,rbstartcol(rb)]);
    maxcol(rb)=max([rbendcol(rb),rbstartcol(rb)-1]);
  end
  for i=1:I
    for j=N(cb):N(cb+1)-1
      newindex(i,j-N(cb)+1)=(i-1)*n(cb)+(n(cb)-
      j+N(cb)-1);
      newindexmod(i,j-N(cb)+1)=mod(newindex(i,j-
      N(cb)+1),n(cb));
      newindexquotient(i,j-N(cb)+1)=fix(newindex(i,j-
      N(cb)+1)/n(cb));
      rowblockindex(i,j-N(cb)+1)=fix(newindex(i,j-
      N(cb)+1)/I)+1;
    end
  end
  for i=1:I
      for j=1:n(cb)
          for rb=1:n(cb)
              if rowblockindex(i,j)==rb;
                 if newindexmod(i,j)>maxcol(rb)
                    col=rbendrow(rb)-
newindexquotient(i,j)+(n(cb)-1-
```

```
newindexmod(i,j))*(rbendrow(rb)-rbstartrow(rb));
                elseif newindexmod(i,j)>=mincol(rb) &
newindexmod(i,j)<=maxcol(rb)
                      if rbstartcol(rb)>rbendcol(rb)
                         c=0;
                         d=-1;
                      else
                         c=1;
                         d=1;
                      end
                      col=(rbendrow(rb)-
rbstartrow(rb))*(n(cb)-1-maxcol(rb))+(rbendrow(rb)-
newindexquotient(i,j)+c)+(maxcol(rb)-
newindexmod(i,j))*(rbendrow(rb)-rbstartrow(rb)+d);
                else %if newindexmod(i,j)<=mincol(rb)
                   col=I-mincol(rb)*(rbendrow(rb)-
rbstartrow(rb))+(rbendrow(rb)-newindexquotient(i,j)
+1)+(mincol(rb)-1-newindexmod(i,j))*(rbendrow(rb)-
rbstartrow(rb));
                end
                   row=1+I-N(cb+1)+rowblockindex(i,j);
             end
          end
                 pr(i,j+N(cb)-1)=row;
                pc(i,j+N(cb)-1)=col;
       end
    end
end

function out=chaoperm(im,pr,pc,num,forward)
%
 [rows,cols] = size(im);
mat = zeros([rows,cols,num+1]);
mat(:,:,1) = im(:,:);
for loc=2:num+1
if(strcmp(forward,'forward'))
  for i=1:rows
    for j=1:cols
      mat(pr(i,j),pc(i,j),loc) = mat(i,j,loc-1);
    end
  end
elseif(strcmp(forward,'backward'))
  for i=1:rows
    for j=1:cols
      mat(i,j,loc) = mat(pr(i,j),pc(i,j),loc-1);
    end
   end
  end
end
out = mat(:,:,num+1);
```

```
function ID=IDMF(x,y)
%this fuunction( irregular Deviation Measuring Factor)
is based on how much the deviation cased by encryption
is
%irregular.
%x:-Original Image
%y:-Encrypted Image
%D:-Maximum Deviation Measuring Factor
x=double(x);
y=double(y);
%first,calculate the difference between each pixel
value before and after
%encryption
D=uint8(abs(x-y));
%calculate the H histogram of the difference
H=imhist(D)
%calculate the average of H
DC=0;
for i=1:256
  DC=DC+H(i);
end
DC=DC/256;
%substract the Dc from H at every point
for i=1:256
  AC(i)=abs(H(i)-DC);
end
%calculate the sum of AC
ID=0;
for i=1:256
  ID=ID+AC(i);
End;
%the lower the value, the better the encryption

function D=MDMF(x,y)
%this fuunction calculates the Maximum Deviation
Measuring Factor which
%calculates the deviation between the original and
encrypterd image.
%x:-Original Image
%y:-Encrypted Image
%D:-Maximum Deviation Measuring Factor
%first,calculate the histogram between the original and
encrypted image
x1=imhist(x);
y1=imhist(y);
%then calculate the difference between the two
diff=abs(y1-x1);
%then calculate D as follows
D1=0;
for i=2:255
```

```
D1=D1+diff(i);
end;
D2=(diff(1)+diff(255))/2;
D=D1+D2;

function y=psnr(im1,im2)
%Peak Signal to Noise Ratio
MSEa=0;
[m,n]=size(im1);
im1=double(im1);
im2=double(im2);
for i=1:m
for j=1:n
x=double(im1(i,j)-im2(i,j));
x1=x^2/(m*n);
MSEa=MSEa+x1;
end;
end;
MSEa=double(MSEa);
y=10*log10((255^2)/MSEa);
```

Index

9780367508159